SAGE was founded in 1965 by Sara Miller McCune to support the dissemination of usable knowledge by publishing innovative and high-quality research and teaching content. Today, we publish over 900 journals, including those of more than 400 learned societies, more than 800 new books per year, and a growing range of library products including archives, data, case studies, reports, and video. SAGE remains majority-owned by our founder, and after Sara's lifetime will become owned by a charitable trust that secures our continued independence.

Los Angeles | London | New Delhi | Singapore | Washington DC | Melbourne

ORGANIC FARMING

ORGANIC FARMING

Economics, Policy and Practices

HARI RAM PRAJAPATI

Los Angeles | London | New Delhi
Singapore | Washington DC | Melbourne

First published in 2020 by

SAGE Publications India Pvt Ltd
B1/I-1 Mohan Cooperative Industrial Area
Mathura Road, New Delhi 110 044, India
www.sagepub.in

SAGE Publications Inc
2455 Teller Road
Thousand Oaks, California 91320, USA

SAGE Publications Ltd
1 Oliver's Yard, 55 City Road
London EC1Y 1SP, United Kingdom

SAGE Publications Asia-Pacific Pte Ltd
18 Cross Street #10-10/11/12
China Square Central
Singapore 048423

Published by Vivek Mehra for SAGE Publications India Pvt Ltd and typeset in 10.5/13 pt Berkeley by AG Infographics, Delhi.

Library of Congress Control Number: 2020943046

ISBN: 978-93-5388-240-2 (HB)

SAGE Team: Abhijit Baroi, Syed Husain Naqvi, Mahira Chadha and Anupama Krishnan

To

My

Late Loving Parents

CONTENTS

DETAILED CONTENTS

Part I: Economic Theories

Part II: Market Structure

LIST OF FIGURES

LIST OF TABLES

LIST OF ABBREVIATIONS

AAGR	Average Annual Growth Rate
AAP	Average Annual Production
AGDP	Agricultural Gross Domestic Product
APEDA	Agricultural and Processed Food Products Export Development Authority
APGMC	Agriculture Produce Grading, Marking and Certification ACT
APIGR	Association for Propagation of Indigenous Genetic Resources
ARISE	Agricultural Renewal Institution for Sustainable Environment
CACP	Commission for Agricultural Costs and Prices
CES	Constant Elasticity of Substitution
COC	Soil Organic Carbon
CLA	Conjugated Linolenic Acid
EFAS	Edinburgh Farming Attitudes Scale
EFIS	Edinburgh Farming Implementation Scale
EFOS	Edinburgh Farming Objectives Scale
FAO	Food and Agriculture Organisation
FCO	Fertilizer Control Order
FTDR	Foreign Trade and Development Regulation Act
FYM	Farm Yard Manure
GGSOC	Participatory Guarantee Systems Organic Council
GOPCA	Gujarat Organic Products Certification Agency (GOPCA)

HMNEHS	Horticulture Mission for North East and Himalayan States
HYV	High Yielding Varieties
ICAR	Indian Council of Agricultural Research
IFOAM	International Federation of Organic Agriculture Movement
IFPRI	International Food Policy Research Institute
IPNM	Integrated Plant Nutrient Management
KVK	Krishi Vigyan Kendra
MKSP	Mahila Kisan Sanshaktikaran Pariyojana
MRS	Marginal Rate of Substitution
MRT	Marginal Rate of Transformation
NAB	National Accreditation Body
NABARD	National Bank for Agriculture and Rural Development
NCDC	National Cooperative Development Corporation
NGO	Non-Governmental Organization
NHM	National Horticulture Mission
NOSB	National Organic Standards Board
NPK	Nitrogen, Phosphorous, Potassium
NPMSHF	National Project on Management of Soil Health and Fertility
NPOF	National Project on Organic Farming
NPOP	National Programme for Organic Production
NMSA	National Mission on Sustainable Agriculture
NRLM	National Rural Livelihood Mission
OAL	Organic Agriculture Land
PSM	Phosphate Solubilizing Micro-organism
PCC	Per Capita Consumption
PGS	Participatory Guarantee System
PKVY	Paramparagat Krishi Vikash Yojana
PROM	Phosphate Rich Organic Manure
RAPI	Rainfed Areas Prioritisation Index
RKVY	Rashtriya Krishi Vikas Yojana
SFAS	Sustainable Agriculture Farming Systems
SRI	System of Rice Intensification
TAL	Total Agriculture Land

TVA	Trans-Vaccenic Acid
UNCTAD	United Nations Conference on Trade and Development
UPASI	United Planter 's Association of South India
USDA	United States Department of Agriculture
WHO	World Health Organization
ZBNF	Zero Budget Natural Farming

FOREWORD

Study of consumer and producer behaviour is central to economic science. This book presents the theoretical, empirical and policy concern of organic farming, an important concern.

The book comprises four sets of ideas: First, *theoretical inquiry*, based on the existing theoretical literature; second, the *market structure of organics*; third, *descriptive analysis* of organic farming in India and Asia; and lastly, *public policy analysis*.

The demand for organic products is high in the market. The agricultural product market structure is non-collusive in nature: There are two market demand curves—one for organic products and the other for conventional products—and two levels of prices. The section on *descriptive analysis* presents state-wise growth of organic farming in India as well as in Asia.

The section on *public policy analysis* tries to present the international organic rules and regulations which are accepted across countries. Certification rules and regulations, initially determined by a third-party certification system and further moving towards public and government body-based participation systems, are important.

This book is recommended to the scholars and practitioners of economics and related areas.

Yoginder K. Alagh
Former Vice-Chancellor, Jawaharlal Nehru University
Former Chairman, Institute of Rural Management Anand
Former Chancellor, Central University of Gujarat

PREFACE

India is one of the largest agrarian economies in the world, where about 44 per cent of the workforce is employed in agriculture, contributing 14 per cent to the GDP and about 10 per cent to the country's exports. However, the productivity of the labour force engaged in agriculture has continuously declined. Conventional farming methods have become unfeasible due to ever-rising input prices. This has led to an increase in rural indebtedness and serious agrarian crisis in India. The methods of conventional farming are not only unsustainable but also impact the environment negatively, deplete natural resources, degrade soil quality, and pollute water and air.

The alternative—low-cost natural farming—is the only solution for debt-ridden farmers. Organic farming or natural farming is the hope for agricultural sustainability. The demand for organic products is ever-increasing and rational consumers are willing to pay higher for organic produces. The world is moving towards organic farming. About 58.7 million hectares of land was converted into organic farmland in 2016. This constitutes 1.2 per cent of the total agricultural land. The Asian region occupies the third place after Oceania and Europe in terms of organic agricultural land and organic output. In the Asian region, India is the leading country and has the largest number of organic producers in the world.

A major challenge at this stage is to ensure quality food and nutritional security for billions of people in India. Currently, the Indian agricultural sector is facing twin challenges which are related to farming methods: One is related with the degradation of natural resources and the other with the cost and quality of agricultural output. The high cost of cultivation is the cause of indebtedness of millions of farmers, and the quality of food and water is the main reason for the increase in many serious diseases. Organic farming may be the solution for the aforementioned challenges and the public policy incentives may be key drivers in the growth of organic farming.

In order to discuss these issues, this book has been divided into four sections: Economic Theories, Market Structure, Descriptive Analysis and Public Policy. A total of 15 chapters are spread across these four sections, covering all possible aspects of organic farming in agricultural economics. The theoretical section includes production theories of economics and household theories of production to explain the basic aspects of organic farming. The market structure section covers the concept of market in economics and dual market possibility within the agriculture market. The descriptive analysis section includes the status and growth of organic farming in Asia and India along with specific Indian states. The public policy section covers the policy provisions for organic farming across the world and in India and its states. It is hoped that this book will be found useful by its readers in understanding the essential aspects of organic farming.

ACKNOWLEDGEMENTS

Writing this book would not have been possible without the guidance of Professor Y. K. Alagh, agricultural economist, and Professor Indira, my doctoral supervisor, along with the consistent moral support of my wife. Also, without the support of my college principal, Dr Kalpana Bhakuni, and the other members of the Department of Economics—Dr Rupa Basu, Dr Sona Mandal, Dr Rupali A. Khanna, Dr Monami Basu, Dr Kh. Pou, Ms Reena Devi, Ms Tanushree Dash, Ms Preeti Mann, Mr Aminesh and Ms Nishtha Sadana—it would have been impossible to complete this work. My close friend Dr Sanjay Marale, an environmental specialist, has guided me from time to time on the issues of natural resources depletion. My sincere gratitude to everyone who strives to grow and helps others grow.

Economic
Theories

I

Organic Producer and Economic Model

INTRODUCTION

The 21st century has brought tremendous changes in agriculture. Through research and development, various advancements have been made in the farming methods during this period. An alternative farming method, which is popularly known as organic farming or organic agriculture, has also been developed. In this method, farming is done without using chemical fertilizers and pesticides. The rationale behind this method is based on four basic principles: Principle of health, principle of care, principle of ecology and principle of fairness. Hence, the main aim of organic farming is production of high-quality food which is rich in nutrients and healthy for human beings as well as nature. The sustainability of any farming method depends on its productivity; therefore, it is important to investigate the productivity and feasibility of organic farming practices. Since it is a new method, a lot of risks are involved in the whole organic movement. Thus, the study on producer behaviour becomes more important than other stakeholders, that is, market and consumer.

The existing conventional farming method uses high-yielding variety (HYV) seeds, chemical fertilizers and pesticides, along with the new technology which increases food production many folds (Bhattacharyya & Chakraborty, 2005). However, at the same time, it creates many health and environmental problems (Giller, Beare, Lavelle, Izac, & Swift, 1997; Singh, 2000). It is a fact that the global food grains production had doubled under conventional agriculture and the world was able to fulfil the food demand of seven plus billion people around the globe (FAO, 2001; Tilman et al., 2001; WHO, 1990). But the food production system ought to meet the twin future challenges: first is the increasing demand for meat and high-calorie diets because of the growing population and second is minimizing its negative impact on environment and its domains, namely, soil, water and air (Foley et al., 2011; Godfray et al., 2010).

In the existing literature, it has been found that conventional farming is not able to meet the second and the most serious future challenge. Therefore, organic farming is finding a place everywhere and has the potential to meet the aforementioned twin challenges simultaneously; first, by increasing productivity by applying scientific technology with indigenous knowledge and second, by minimizing the impact on health and environment (Graef, Schütte, Winkel, Teichmann, & Mertens, 2010; McIntyre, Herren, Wakhungu, & Watson 2009; Seufert, Ramankutty, & Foley, 2012). The worldwide organic movement creates awareness among consumers and producers across countries. As a result, both are interested in consuming organic quality food and prefer to enjoy a healthy life. Health and environment are two commonly stated reasons for consumers' preference for purchasing organic products (Wandel & Bugge, 1997). Health is a major concern, according to many consumers for the purchase of organic food (Land, 1998; Magnusson et al., 2001; Wandel & Bugge, 1997) and environment may be for some others (Schifferstein & Ophuis, 1998). The number of consumers is continuously increasing in this group which boosts the demand for organic food, resulting in more and more adoption of organic farming.

There is a big debate on productivity issue of both farming methods and social scientists are divided into two groups based on their opinions. The ones who support organic farming have presented their facts related to increasing productivity and health of the ecosystem (Badgley et al., 2007). The others who oppose organic farming have argued that it may contain lower yield and, therefore, needs additional land to produce the same amount of food, leading to deforestation and biodiversity loss, and thus undermining the environmental benefits of organic practices (Cassman, 2007; Connor, 2008; Trewavas, 2001). Making the counterargument to the critics of organic farming, various individuals and meta-analysis studies have compared the yields of conventional and organic farms. These studies have found that organic farms have lower yields than conventional farms, but the yield differences are highly contextual and depend on the structure and characteristics of the farm. A study conducted by Seufert et al. (2012), based on meta-regression analysis, has found that the range of organic yields is 5 per cent lower than the conventional yields in rainfed regions and 13 per cent lower in fertile regions. Thus, this result suggested that organic farming initially begins in the lowest yield gap rainfed regions.

The critics of organic farming have always questioned that it has lower productivity, but a young farmer Sumant Kumar, of Darveshpura village of Nalanda district in Bihar state, has made a world record in rice production through organic farming, using the system of rice intensification (SRI). He has grown an incredible 22.4 tons/ha of rice, surpassing the previous 2011 world record of 19.4 tons/ha of Chinese scientist Yuan Longping, known as the 'father of hybrid rice'. This yield is not just a one-time thing and limited to Sumant alone, but his neighbour farmer in the same village has recorded over 17 tons/ha, double than his usual yields. Another farmer Nitish Kumar of the same Darveshpura village has obtained the world record with 72.9 ton/ha production of potato via the participatory method of organic farming in the period 2011–2012. After that, Rakesh Kumar at Sohdih village in the same district of Bihar is another farmer who had grabbed the world record for potato production, with a yield of 108.8 tons/hectare in the year 2013 using only vermicompost for vegetable production. These farmers' achievements inspired Nobel laureate economist Joseph Stiglitz and he visited the Nalanda district in 2013, and recognized the potential of this kind of organic farming. He told the villagers that they were 'better than scientists' (Ojha & Saha, 2016). These significant achievements of few individual farmers have shown the potential of organic farming and have busted all the myths related to lower productivity.

In India, organic farming movement started in the 2000s and made a remarkable progress in the production of organic input and output. The country has a comparative advantage in organic farming because nearly half of the cultivated area is rainfed, constitutes more than 55 per cent of the net sown area of the country and is home to two-third livestock and 40 per cent of human population. To gain from this opportunity, India had started two programmes, National Programme for Organic Production (NPOP) in the year 2000 and National Project on Organic Farming (NPOF) in the year 2004. With the classification of three purity zone areas: Category—I, Category—II and Category—III, the results of these programmes are appreciable. In the year 2000, the total certified area, including wild collection under organic farming was 2,775 hectares which increased many folds and recorded 5.71 million hectares in 2015–2016 (APEDA, 2016).

The growth of certified organic area varies from state to state in India. This variation has many causes, namely state's policy on organic farming, niche market and implementation of programmes on organic farming, as well as the central government projects, input market avail-ability, awareness and information related to organic farming among producers. Madhya Pradesh has acquired the first place, having 1.1 million hectares of its farming area under organic farming, which is nearly 52 per cent of the total certified organic area. Maharashtra is at the second position and covers 0.96 million hectares (33.6%) area, while Orissa is at the third position with 0.67 million hectares (9.7%) of its farming areas. Uttarakhand and Sikkim are organic states. The position of Gujarat is satisfactory, and farmers are growing crops such as cotton, cumin, cereals, spices, vegetables and fruits using organic farming processes (Mitra & Devi, 2016).

The focus of research in Gujarat has been on four rainfed districts, namely, Ahmedabad, Banaskantha, Patan and Surendranagar for the field survey, which fall in the same agro-climatic zone. How to attain long-run sustainable agriculture growth is the major challenge for the policy makers and the government. Although the Government of Gujarat had announced the new organic farming policy in July 2015, how effective it was and was it able to promote organic farming are questions of further research. Sustainable agriculture growth is pos-sible through promoting organic farming in rainfed and conventional agriculture irrigated regions simultaneously because it is now proven by various experimental and field studies that there are very less differ-ences in yield under both the systems in rainfed region. Thus, Gujarat has an opportunity to maintain the double-digit agriculture growth rate in future by providing alternative low-cost farming system. This low-cost system reduces the use of chemical fertilizers, ground water for irrigation and hybrid seeds, which are big threat to sustainable agriculture in the state.

MEANING AND DEFINITIONS

In economics, producer is defined as 'a person who creates economic value'. It means that a producer is someone who adds value in goods and services by combining, land, labour and capital. However, a producer can also be stated as 'a person who puts together the factors of production in order to produce goods and services for commercial use'. The above definitions keep in mind that a producer may produce goods or services for his/her own consumption and not always for sale. However, in economics the producer is defined in a rigid sense, a person may not be called a producer if he/she has little or no impact on the market or if he/she has not enough to sale. For example, a subsistence farmer produces crops for self-consumption and not for sale, Hence, he/she may not be considered as a producer in an economic environment even though he/she in fact is producing something. This is because economic analysis is only concentrated on buying and selling of goods and services, and the exchanges thereof, rather than just consumption and production. In economics a person involved in production and selling activity is called a 'firm' and usually economists have this in mind in the study of producer behaviour.

The criteria of classifying a person as a producer is not so clear in economics; it may be an individual, farm household, firm, company and government unit and so on. It depends on the objective of research and what exactly the researcher wants to study in a particular context. In this book, producer means 'farm household' that is producing crops for both consumption and selling purposes. The goal of farm household is to produce crops at a minimum cost with a given alternative technology and earn profit by selling a part of the produce. The choice of alternative production technology studied under the theory of producer behaviour in economics provides a hypothetical framework under complete knowledge axiom. But like other decision-makers, farm households rarely have full information related to weather, monsoon, climate, rainfall and soil contents, and so on. Therefore, he/ she has the following choices: the optimal allocation of resources with the available farming method, crop selection, use of input and so on.

NPOP has explained the meaning of organic agriculture as, 'A system of farm design and management to create an agro-ecosystem, which can achieve sustainable productivity without the use of artificial external inputs such as chemical fertilizers and pesticides'. According to this idea, agro-ecosystem functions optimally if the diversity in crops and animal husbandry is managed in such a manner that improves producer welfare and soil quality. Different individuals and institutions have also defined organic farming in their own ways. Institutions like the International Federation of Organic Agriculture Movements (IFOAM) in 1972 defined, 'Organic agriculture is a production system that sustains the health of soils, ecosystems, and people. It relies on ecological processes, biodiversity and cycles adapted to local conditions, rather than the use of inputs with adverse effects.' The United States Department of Agriculture (USDA) and National Organic Standards Board (NOSB) in 1995, defined organic farming as 'an ecological production management system that promotes and enhances biodiversity, biological cycles, and soil biologi-cal activity. It is based on the minimal use of off-farm inputs and on management practices that restores, maintain and enhances ecological harmony.' Individuals such as Martin (2009) also defined, 'Organic farming is a method of crop and livestock production that involves much more than choosing not to use pesticides, fertilizers, genetically modified organisms, antibiotics and growth hormones.' Yadav et al. (2013) defined organic farming as 'a production system which avoids, or largely excludes, the use of synthetic fertilizers, pesticides, growth regulators, and livestock feed additives.'

THE ECONOMIC MODEL

In microeconomics, producer behaviour is based on the economic theory of firm's profit maximization hypothesis. However, this hypoth-esis and its methodological foundation have been criticized on many grounds. In agriculture production activity, consider a farm household as producer. The profit maximization assumption is substantially more difficult than cost minimization because in agriculture, risk-related problem is always involved. In economic literature investigations of producer's cultivation behaviour, most economists applied net profit analysis and calculated it by total revenue (TR) minus total cost (TC). Here, TR is the total income received from sale of output and TC is the total expenditure made in cultivation. This approach is more popular and frequently applied in empirical investigation. But after the invention of the production function approach and advances in econometric techniques, the whole research on production economics has spread in different directions. Some economists are interested in deriving TC curves and others the calculation of factor elasticity, the elasticity of substitution and technical efficiency with the help of a production function. But the use of a production function for estimation of parameters has several problems, such as, homogeneity of degree, the price being an exogenous variable, the problem of multi-collinearity, estimation of elasticity and economics of scale. For eliminating these problems, the dual approach was invented. The dual formation of cost curves through production and cost functions was established by Shephard in 1953, also popularly known as duality theory of production. This new formulation of duality theory helped to overcome the limitation of traditional production theory. The detailed description related to duality theory and its application is presented in Chapter 4.

Theory of production of firm explains that the producer is motivated by the desire to maximize profit, which is made by choosing appropriate technology that can minimize the cost of production or maximize profit. A similar approach is followed in the investigation of farmer's production behaviour analysis. But in the case of farming business, the individual farmer does not take any decision related to the adoption of a particular technology, purchase of machinery or other inputs. Generally, the decisions are taken by the whole family collectively. Hence the word 'farm household' is more appropriate than farmer, who is also called a producer. The production technology of a farm household is influenced by its own consumption needs as well as market demand along with the availability of labour, land and other inputs. Any changes in its production technology depend on the demand of product and price. As found in the literature, the demand for organic products is very high in both the domestic and international markets (Oberholtzer, Dimitri, & Jaenicke, 2013). It provides an opportunity to producer to explore the market demand and receive gains from trade.

A large number of studies have modelled farmers' decision behaviour in the line of the classical theory of firm by assuming that profit maximization is the only objective for the motivation to adopt new technology. Accordingly, they all ignored the reality that decisions of farm household are usually influenced by multiple goals rather than profit maximization only. Farm households also face a financial problem: trade-off between investment in farm and family consumption expenditure. The latter is likely to be considered. In such a situation, the decision-maker is generally seeking an optimal cooperation among several objectives and trying to achieve a satisfying level of his goal (Romero & Rehman, 1989). Thus, farm household is interested in managing multiple goals in such a way that increases gross profit, reduces indebtedness, avoids risk, increases productivity, maintains land quality, improves family living standard and there is enough time for leisure and so on.

The decision function of the farmers would become simplified in a world free of risk and uncertainty (Walker, Heady, Tweeten, & Pesek, 1960). The static economic theory stated that the decision-maker has complete knowledge. Like other decision-makers, farm households rarely have complete information about weather, monsoon, climate, rainfall and soil contents. Thus, farmers' decisions are involved in the choice of farming method, crop selection, use of input and so on. Some uncertainty will always be present in their decisions. In case of choice of crops, their past experiences on yield influence their present decisions. Minimizing the gap between expected and actual yield is the normative approach of decision-making, which is a problem of descriptive economics.

Price and Output Determination in Non-collusive Market

In production economics, different types of market structures exist such as monopoly, imperfect competition and perfect competition under which firms, cost, production, input purchasing and output selling behaviour are studied. But in real life, oligopoly market exists, where the number of the producers are limited and they produce differentiated products. Though the application different market structure or condition are found in agricultural food demand and supply market and also for input market structure, 'farm household' production behaviour, is a crucial task. Price and cost play a crucial role in choosing product technology. The prices of food products are determined by the demand and supply schedule of the product type. While the output is determined by the area and number of producers engaged in producing with the chosen farming method. The prices of organic products are higher than common products due to the higher demand in niche, national and international markets because of the supply constraints. In India, the total certified land under organic farming is less than 1 per cent of the total cultivated area, therefore, the production or supply is much less than the demand, and hence the prices are high. Another reason for higher prices is lower per hectare productivity. Additional certification and inspection costs make it costlier than common products as well.

Consider a closed economy in which agriculture sector has two types of farmers, practicing organic and conventional farming, with non-collusive oligopoly market structure, producing differentiated products using different production technology. The first producing organic products and targeting an urban niche market and the second producing a common product devoted to a large national market. Both have an opportunity to sell their products in the local, urban and the national markets. The adoption of organic farming occurs in case farmers get higher returns than conventional farmers or are able to produce at a lower cost. It means the producers are motivated by profit only. Within this structure, if the total number of farmers are N, then each farmer will choose either organic farming or conventional. The number of organic farmers denoted by n (small group), remaining $(N - n)$ is conventional growers (large group).

I have followed Chamberlin's small group model (1933) and Sweezy (1839) kinked demand curve model for the two groups of farmers. The conventional grower can sell their product with price P_c in the national market and capture a major part of the national food

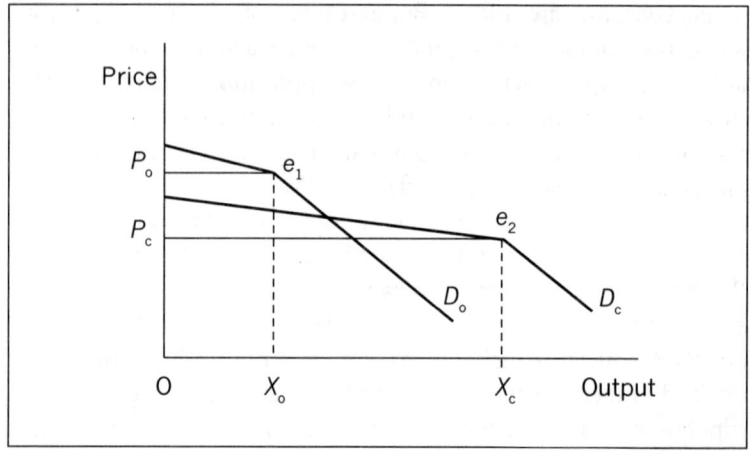

Figure 1.1 *Price and Output Determination in Two Groups' Model*

grain market. Other organic growers have an 'urban niche market' to sell their products at a price of P_o which is generally higher than P_c. Here, it is assumed that market demand is facing the dual price and output problems. The price varies due to the variation in the cost of production and output due to difference in technology.

ISSUE OF ORGANIC PRODUCER AND ECONOMIC THEORY

The economic literature survey conducted by Crowder and Reganold (2015) and suggested that organic producer earn profit and knowing the cause of the shift towards organic farming are based on the comparison of conventional and organic farming, and focused on the empirical analysis; few studies have tried to explain theoretically. There is no evidence of two group models in literature that can theoretically explain the price and output determination for both the markets. The multiple goals of the farmers are not examined in the case of organic farmers. The formulation of a new economic model is essential to consider the multiple goals of farmers rather than a single goal. It is important to analyse how farmers behave in case of non-collusive duopoly domestic market.

Analysing the impact of both conventional and organic farming systems on environmental domains is also necessary for knowing the extent of environmental pollution problem. This study tries to integrate both theoretical and empirical investigation, and will be fruitful for policy formulation on organic farming that will give justice to both producers and consumers, along with the environment.

The Government of India and the Ministry of Agriculture have made tremendous efforts in this regard and specified the areas for the promotion of organic farming across states according to the Planning Commission's (2012) report titled *Prioritisation of Rainfed Areas in India*, which is based on the available data set of 499 districts across the country. The regions not included in the report are Jammu and Kashmir, districts of the North-eastern region except Assam, districts of Goa and the union territories. These 499 districts account for more than 90 per cent population and area of the country. One-third (167) of the total 499 districts are considered as high-priority rainfed districts based on Rainfed Areas Prioritization Index (RAPI) score. It was calculated by the crop and livestock-based interventions. In this categorization scale, half of the districts (13) of Gujarat fall under the rainfed categories.

However, a large number of farmers who live in the remote and rainfed areas are still practicing traditional livestock-based organic farming system. The average use of chemical fertilizers was recorded as 18.5kg in non-irrigated districts, while it was 58kg in the irrigated districts—more than three times (Katyal & Reddy, 1997). The estimation made by different individuals based on secondary data sources recorded that nearly 30 per cent of farmers in rainfed areas do not use chemical fertilizers and pesticides in India. Thus, several farmers are practicing organic farming by default. For investigation of this issue, the Government of India formed a task force on organic farming in 2001. It identified that the rainfed districts and the North-eastern states have enough potential for organic farming due to the use of low chemical fertilizers used and water scarcity. Several other reviews and research also support this view (Dwivedi, 2005; Gopinath et al., 2011; Ramesh, Singh, & Rao, 2005).

The adoption of organic farming in rainfed districts of the country will be gainful for the producers, consumers, environment and the nation. The utilization of dry land and arid farms by adopting horticulture, livestock rearing and growing vegetables under organic farming is profitable for the individuals and the country. So the research problem is crucial for individual, society, environment and nation. The studies related to farmers' decision-making are found in other developed and developing countries, but some studies were also conducted by the International Food Policy Research Institute (IFPRI) in Tamil Nadu. In this direction a comparative study was made by Puttaswamaiah for Gujarat and Karnataka.

Thus, this book helps to understand the farm household's decision-making behaviour to adopt new organic farming methods and resource allocation thereof. Simultaneously, it provides the evidence of growth of organic farming in India as well as across the countries. The critics of organic farming have raised the issue of productivity gap, but that can be increased by using best organic seeds and modern farming tools. There are two types of farmers who adopt organic farming: one who adopt it for making profit by producing cash crops and others for producing food grains, vegetables, fruits and so on. So the basic questions which are investigated in this book are: What are the factors that influence farm households to adopt organic farming in India and other countries across the world? Which farming system is more economically efficient and maintains long-run sustainability? What are the environmental impacts of using these two farming systems?

The objective of this book is to present theoretical, practical and policy-related challenges organic producers face. In the theoretical and market structure section we examine the economic model in case of two groups, two commodities and two market models. The empirical results of producers' adoption behaviour of organic farming and the factors that influence farm households to adopt organic farming in India and other countries across the world are presented in the book. In the analysis section we examine application of the model along with description of status of organic farming across states in India as well India as a whole; the public policy related to organic farming falls under the public policy section.

CONCLUSION

This chapter presents an overview of the entire book, which is used for the economics analysis of the producer behaviour. The study on organic farming has different dimensions such as the issue of producer, consumer, product and market. Each stakeholder has their own problem; this book mainly concentrates on the producer and market. The economic theory on producer behaviour is related to the problem of production economics, which comes under microeconomics and is studied using different types of market structures. The aim of the producer is always the same in all different kinds of markets, but the quality of the product and the number of producers changes from market to market.

REFERENCES

Badgley, C., Moghtader, J., Quintero, E., Zakem, E., Chappell, M. J., Aviles-Vazquez, K., … Perfecto, I. (2007). Organic agriculture and the global food supply. *Renewable Agriculture and Food Systems*, 22(2), 86–108.

Bhattacharyya, P., & Chakraborty, G. (2005). Current status of organic farming in India and other countries. *Indian Journal of Fertilisers*, 1(9), 111–123.

Cassman, K. (2007). Editorial response by Kenneth Cassman: Can organic agriculture feed the world-science to the rescue? *Renewable Agriculture and Food Systems*, 22(2), 83–84.

Chamberlin, E. (1933). *Theory of Monopolistic Competition*. Harvard University Press Cambridge, pp. 10–213

Connor, D. J. (2008). Organic agriculture cannot feed the world. *Field Crops Research*, 106(2), 187–190.

Crowder, David W., and John P. Reganold (2015). Financial competitiveness of organic agriculture on a global scale. *Proceedings of the National Academy of Sciences*, 112(4), 7611–7616.

Dwivedi, V. (2005, 10–11 March). *Organic farming: Policy initiatives* (pp. 58–61). Paper presented at the National Seminar on National Policy on Promoting Organic Farming, Ghaziabad.

Foley, J. A., Ramankutty, N., Brauman, K. A., Cassidy, E. S., Gerber, J. S., Johnston, M., … & Balzer, C. (2011). Solutions for a cultivated planet. *Nature*, 478(7369), 337.

Food and Agriculture Organization (FAO). (2001). *The state of food insecurity in the world*. Rome: Author.

Giller, K. E., Beare, M. H., Lavelle, P., Izac, A. M., & Swift, M. J. (1997). Agricultural intensification, soil biodiversity and agro ecosystem function. *Applied Soil Ecology*, 6(1), 3–16.

Godfray, H. C. J., Beddington, J. R., Crute, I. R., Haddad, L., Lawrence, D., Muir, J. F., … Toulmin, C. (2010). Food security: The challenge of feeding 9 billion people. *Science, 327*(5967), 812–818.

Gopinath, K. A., Venkateswarlu, B., Venkateswarlu, S., Yadav, S. K., Balloli, S. S., Srinivasa Rao, Ch.,… Maheswari, M. (2011). *Organic sesame production* (Technical bulletin, p. 34). Santoshnagar: Central Research Institute for Dryland Agriculture.

Graef, F., Schütte, G., Winkel, B., Teichmann, H., & Mertens, M. (2010). Scale implications for environmental risk assessment and monitoring of the cultivation of genetically modified herbicide-resistant sugar beet: A review. *Living Reviews Landscape Research, 4*(3). Retrieved from http://lrlr.landscapeonline. de/Articles/lrlr-2010-3/download/lrlr-2010-3BW.pdf

Katyal, J. C., & Reddy, K. C. K (1997). Plant nutrient supply needs: Rainfed food crops. In J. S. Kanwar & J. C. Katyal (Eds.), *Plant nutrient needs, supply, efficiency and policy issues:* 2000–2025 (pp. 91–113). New Delhi: National Academy of Agricultural Sciences.

Land, B. (1998). *Consumers' dietary patterns and desires for change* (Working Paper No. 31). Roskilde: Roskilde University.

Magnusson, M. K., Arvola, A., Koivisto Hursti, U. K., Åberg, L., & Sjödén, P. O. (2001). Attitudes towards organic foods among Swedish consumers. *British Food Journal, 103*(3), 209–227.

Martin, H. (2009). Introduction to organic farming, order no 09-077, factsheet 06-103 APEDA (2016) Anual Report 2015–16.

McIntyre, B. D., Herren, H. R., Wakhungu, J., & Watson, R. T. (2009). *Agriculture at a crossroads synthesis report.* Washington, DC: International Assessment of Agricultural Knowledge, Science and Technology for Development (IAASTD).

Mitra, S., & Devi, H. (2016). Organic Horticulture in India. *Horticulturae, 2*(4), 17.

Oberholtzer, L., Dimitri, C., & Jaenicke, E. C. (2013). International trade of organic food: Evidence of US imports. *Renewable Agriculture and Food Systems, 28*(3), 255–262.

Ojha, M. D., & Saha, B. (2016). Organic Potato in Nalanda (Bihar): Using eco-friendly agri. bios inputs. *Indian Research Journal of Extension Education, 14*(3), 119–121.

Planning Commission (2012). Prioritization of Rainfed Areas in India, Study Report 4, NRAA, New Delhi, India, 100p.

Ramesh, P., Singh, M., & Rao, A. S. (2005). Organic farming: Its relevance to the Indian context. *Current Science, 88*(4), 561–568.

Romero, C., & Rehman, T. (1989). *Developments in agricultural economics: Multiple criteria analysis for agricultural decisions.* Netherlands: Elsevier.

Schifferstein, H. N., & Ophuis, P. A. O. (1998). Health-related determinants of organic food consumption in the Netherlands. *Food Quality and Preference, 9*(3), 119–133.

Seufert, V., Ramankutty, N., & Foley, J. A. (2012). Comparing the yields of organic and conventional agriculture. *Nature, 485*(7397), 229–232.

Singh, R. B. (2000). Environmental consequences of agricultural development: a case study from the Green Revolution state of Haryana, India. *Agriculture, Ecosystems & Environment, 82*(1), 97–103.

Sweezy, P. M. (1939). Demand under conditions of oligopoly. *The Journal of Political Economy, 47*(4), 568–573.

Tilman, D., Fargione, J., Wolff, B., D'Antonio, C., Dobson, A., Howarth, R., … Swackhamer, D. (2001). Forecasting agriculturally driven global environmental change. *Science, 292*(5515), 281–284.

Trewavas, A. (2001). Urban myths of organic farming. *Nature, 410*(6827), 409–410.

Walker, O. L., Heady, E. O., Tweeten, L. G., & Pesek, J. T. (1960). Application of game theory models to decisions on farm practices and resource use. *Research Bulletin (Iowa Agriculture and Home Economics Experiment Station), 33*(488), 1.

Wandel, M., & Bugge, A. (1997). Environmental concern in consumer evaluation of food quality. *Food Quality and Preference, 8*(1), 19–26.

World Health Organization (WHO) (1990). *Public health impact of pesticides used in agriculture.* Geneva: Author.

Yadav, S. K., Babu, S., Yadav, M. K., Singh, K., Yadav, G. S., & Pal, S. (2013). A review of organic farming for sustainable agriculture in northern India. *International Journal of Agronomy.* doi: 10.1155/2013/718145

Economic Theories and Organic Farming

INTRODUCTION

Organic farming is growing as a branch of agricultural economics, so theories which are used in literature to investigate the problem of production economics are similar. Each branch of social science has made tremendous progress in analysing the problems of the related area. Similarly, different advances have been made in the area of production economics and analyses of producer behaviour. The decision of an organic producer to produce an organic product or not is similar to the decision of other producers or firms, which are theoretically examined in the domain of agriculture and production economics. Here the decision behaviour of an organic producer to grow crops organically or not is examined under the theory of production.

The study of an individual firm or producer falls within the area of microeconomics, where a producer makes a decision on his profit and this profit is estimated either using cost or profit function using prices of inputs. Here the same methodology has been followed to study the behaviour of an organic producer because the aim of an organic producer is also profit maximization, which is the same as the goal of firm. There are two most popular approaches in economics which are applied to study the profit of firm or producer, one is 'profit function approach' under which TR and TC are used to calculated net profit. The other is 'cost function approach' under which TC, average cost (AC) and marginal cost (MC) are used.

This chapter presents production economic theories and their empirical application in organic farming. A producer's decision in case of organic farming choice involves economic comparison of farming system, profitability, impact on environmental domains and health issues of conventional and organic foods. Here, theoretical investigation presented historical development of production economics theories. At the same time, empirical application of these theories is also investigated in case of organic farming to find out productivity gap, cost difference and differences in quality of products.

This chapter has two sections. The first section presents the economics theories which can help to explain organic farming and producer behaviour based on production function approaches. The next section deals with the application part of these theories in case of agriculture and related organic farming activity.

The development of production economics theories began with the classical era known as the classical theories of production, where any change in the price of factors of production and its impacts on TC and output of production were studied. In this era there were only three factors of production, namely land, labour and capital. In this phase economists assumed that technology remained constant during the production process. The second and crucial phase was the neoclassical era known as the neoclassical theories of production, where the different types of production function were invented and mathematical techniques applied to study the relationship between input and output. The latter phase (the 1950s) was the modern era known as modern theories of production, where the duality approach was invented and applied to investigate the profit-making behaviour of a producer or firm. Following is a review of theoretical advances and empirical applications in enquiry of organic farming.

CLASSICAL THEORIES OF PRODUCTION AND ORGANIC FARMING

The era of classical phase was the early stage of development where the society was purely dependent on natural resources. So the theories which were propounded were mainly based on physical factors of production and the role of technology was negligible. The relation between inputs and output was examined with the help of a simple production function. The aim of a producer or firm was to maximize yields subject to physical factors of production. In this era, the cultivation of crops was traditional, so there was no need to worry about depletion of natural resources. But these production economic theories were applied for the economic analysis of organic farming presently not consider depletion of natural resources..

In production economic analysis the invention of production function was a turning point when fundamental changes were made such as to formulate a function relationship between inputs demand and output supply. But the role of technology remained constant in the formulation of functional relationship between inputs and output. The classical economists describe production as a function of economic environment and constant technology. In agricultural economics, technology means farming method which alters the use of inputs. This alternative production function investigation begins to apply empirical economic data of inputs demanded, output supplied, price and cost to calculate profit. The advantage of invention of production function was that it helped to establish a causal relationship between output and inputs that lead to theoretical development. Although the classical economists held several assumptions to formulate classical theory of production with the help of production function, these assumption helped in formulating theories and avoiding computational problem in the analysis of production. But the invention of production function open door to develop more complex production function (Tinbergen, Jorgenson, & Waelbroeck, 1978).

The seminal work on factors returns by Von Liebig (1855) was primarily found that if factors of production are increased continuously, then after a certain extent the returns of factors begin to decline in agriculture. It means there are certain other factors of production like climate, sunlight and water, which have minimum requirement to growth of plant. This crucial finding lead him to propound the law of minimum in agronomy. His finding specified that the yield of crop is dependent on the supply of inputs, which are available in smaller amount but are essential for the growth of a plant. The above, 'the law of the minimum', can be also applied in organic farming during the conversion period when the use of organic inputs increases the organic output slowly. The yield of organic output also follows the law of the minimum.

Furthermore, the cost function approach was developed to analyse the input–output relationship, assuming that production coefficients are fixed. For the illustration of functional relationship, mathematical cost equations were used. Later the application of these cost equations in economic theory was found in the work of L. Walras (1876). He made a strong assumption to formulate cost equations, that is, a unit of time was used to produce commodities. He also classified factor services into three categories: services of labour (L), services of land (N) and services of capital (K). The respective prices of commodities were denoted with price and related commodity subscript and similarly was done for respective prices of the services. This method of estimation contained many identical cost equations for prices of unknown commodities. The production coefficients of respective commodities and marginal productivity of inputs are calculated using this method (Walras, 1900). The estimation of production coefficient under the above process was criticized by Schultz who put an argument that some production coefficients may be constant and at the same time other services can vary with change in production of a commodity, some services may be increased and others may be declined (Schultz, 1929).

The aim of the classical theory of production was profit maximization at given cost and revenue constraints. For the calculation of profit, total approach was most popular where profit was calculated by subtracting TC from TR earn from selling output. The short run average variable cost (AVC) and marginal cost (MC) are frequently used in economics textbook, which was propounded by Viner (1931) and Cassels (1936). They laid the foundation for the two-factor model in production economics under the assumption of perfect competition in the market. It is popular that Cassels's theory of production helped to develop Viner cost curves (Larson, 1991). But this view was criticized by Krauthamer (1963), Maxwell (1965) and Danø (1966) on the basis of assumptions which are assumed for construction of short-run AVC and MC form traditional production function. The short-run AVC and MC curves are constant up to a critical level of output then rise upwards, forming a J-shaped curve. Maxwell further argued that whether the shape of a cost curve will be 'U' or 'J' purely depends on the type of capital used in production which is either fixed or variable.

The classical production theory is based on logical reasoning and axioms. In the 1890s, the most debated problem was that production coefficients were treated as constant or variable. The economists were divided into two groups. The ones who supported that coefficients were fixed were Pareto, Barone and others. Another group that supported variability of coefficients consisted Wicksteed, Marshall, Wicksell and Walras. The judgement of later group was universal and used for all the sectors of economy everywhere, while the judgement of the former was limited and rejected because it had some issues in long-run equilibrium analysis. The neoclassical economist Stigler (1939) tested the variable coefficient hypothesis for short-run problem of production and distribution.

The classical economists, Walras in 1874, Pareto in 1927 and Geogescu-Roegen in 1935 were unable to resolve the issue of production coefficient rigidity. Later Stigler tested and proved that production coefficients are variable in nature for short run period. He used the law of demonising variable proportion and marginal productivity of factors to explain the changes in output with respect to change in one factor of production while others remaining constant. This production coefficient variability and rigidity can be also tested for organic agriculture production and has scope for further research. Organic farming is part of the agriculture sector and the above classical economic theory is applied to the agriculture sector. So the above historical development of production economic theory is the part of agricultural economics and rooted to organic farming by default.

NEOCLASSICAL THEORIES OF PRODUCTION AND ORGANIC FARMING

The neoclassical economists tried to resolve the unsolved issues of the classical production theory. Hicks (1946) wrote a book *Value and Capital*, followed by Samuelson (1947) who wrote a book *Foundation of Economic Analysis*. Both of them applied the production function for profit maximizing behaviour of firm or producer. They also used cost as a single restriction for producer choice under demand and supply function. They used implicit function[1] for input and output relationship along with production function. These tools can be also applied in the analysis of organic producer behaviour analysis.

The use of implicit functions for demand and supply with respect to relative prices of inputs and output has some difficulties in econometric estimation. So formulation of a specific econometric model to study producer behaviour based on explicit function of demand and supply must be tested. These functions are used to get the value of parameters from empirical data which helps to measure substitution of inputs, elasticity of input and output price, technical change and economics of scale. The classical approach of producer behaviour is based on assumption of homogeneity and additive of production function. The supply and demand function are derived from production function at equilibrium level of output through profit maximization. In this approach imposing constraints on production pattern poses some empirical difficulty (Jorgenson, 1986).

The invention of production function made incredible improvement in production analysis and its credit goes to Wicksell who developed its functional form in the 19th century. The early empirical work using production function was conducted by Cobb and Douglas (1928), which was based on 23 years of data of the US manufacturing sector during 1899–1922. Both tried to explain the functional relationship between output and inputs with the help of empirical results. Further work on production function was conducted by Brown (1966), Sandelin (1976) and Samuelson (1979).

Cobb and Douglas (1928) have made remarkable contribution by constructing mathematical functional form of production function and measuring volume of output in the US manufacturing sector. They tackled two core issues of the production problem: first, to measure the change in quantity of output due to change in quantity of factor of production such as labour (L) and capital (K). Second, to determine the

---

[1] The implicit function theorem is a device for solving equations, and these equations can have many different settings; see Krantz and Parks (2013).

relationship between output (Q) and two inputs labour (L) and capital (K). The Cobb–Douglas production function is written as follows:

$$Q' = A\, L^{\alpha}\, K^{1-\alpha}, \qquad (2.1)$$

where Q' shows the actual production of Q and Q' is the approximation of Q over a time period. Q' approaches zero as either L or K approaches zero. The Q' is derived from Q, which is independent from Q' and Q' is closely correlated with Q' when incorporated secular trend.

Further, Dhrymes (1965) estimated the degree of homogeneity using the similar production function as $w = A\, Q^{\beta}\, L^{\gamma}$. Where w is the monetary wage rate of labour, L is the amount of labour used and Q is the output. This equation can also be derived from the CES production function and the parameters are estimated with the help of logarithms. The degree of homogeneity of factors of production is calculated as $h = (1 + \gamma)/(1 - \beta)$. This homogeneity estimation suffers the same problem as mentioned above. Dhrymes derived the labour share equation $\alpha_t = (\omega_t\, L_t)/Q_t$ with a constant share of labour in an economy. This share equation can also be written as $w_t = \alpha\, Q_t L_t^{-1}$, which is also called the Dhrymes equation. The regression result of log of the above equation coefficient must be $\beta = 1$ and $\gamma = -1$. When the share of labour is huge, the variation result of regression will be weak.

In 1967 Douglas published a paper commenting on the Cobb–Douglas production function. He commented on both its structure and empirical results. He used the same 23 years of manufacturing sector data on output, labour and use of capital in USA, during 1899–1922. He took log of output, capital and labour and computed capital output efficiency, where capital was deflated to constant purchasing power of dollar. He plotted three curves using logarithmic value of output, labour and capital. The result was surprising that the distance between output curve and labour curve was recorded one third to one quarter. The growth rate of labour curve was noted to be the lowest, while the capital curve was recorded as the highest. The variation of growth rates was recorded overs the years. During the depression years, growth of output and labour declined but capital remained unchanged.

In 1979, Samuelson constructed a hypothesis on the Cobb–Douglas production function as regression of aggregate Cobb–Douglas function held all value added accounting identity. The above hypothesis means that the value added of output is equal to the sum of bill of wages and total profit. After this claim, Schumpeter was surprised and commented that now further technical progress of the Cobb–Douglas function would not be allowed and left a note that Douglas's follower might wish to derive above proof several times. Bowley stated the same relative wages for the United Kingdom and USA. But these facts were not helpful to increase the acceptances of the aggregate neoclassical-type Cobb–Douglas production function.[2]

Felipe and Adams (2005) empirically confirmed Samuelson's claim that regression result of aggregate Cobb–Douglas type production function reproduced the value added income or output accounting identity. Their result supported the Samuelson additive output–input hypothesis. They also concluded that it can be applied to macroeconomics policy formulation. Their empirical finding established that neither aggregate production function nor neoclassical hypothesis have constant returns to scale in competitive market.

Further in 1971, Fisher used simulation analysis to estimate aggregate Cobb–Douglas production function and his statistical results were very close. Though the data used for regression analysis was purposely violated, he reported that in some cases, the constant share of factors of production increases and vice versa also. He concluded, 'consistency in the share of labour is due to use of aggregate type Cobb–Douglas type production function and its success is due to relative consistency in share of labour' (Fisher, 1971).

The use of Cobb–Douglas production function in production analysis began with their own work in 1928 and further applied in empirical research for nearly two decades (see also Douglas 1948, 1967, 1976; Heady & Dillon, 1961; Samuelson, 1979; Walters, 1963). The limitations of Cobb–Douglas production function were pointed out by Arrow, Chenery, Minhas and Solow jointly in 1961, which

[2] Paul Douglas's 'Measurement of production functions and marginal productivity by Samuelson' (1979, p. 931).

was popularly known as the ACMS approach. They (Arrow, Chenry, Minhas and Solow) have noticed that in using the Cobb–Douglas production function and prior imposing restriction on pattern of inputs substitution. They added a constraint, that is, elasticity of substitution of all inputs must be equal to one, which resulted in the constant elasticity substitution (CES) production function. Later they added flexibility in use of inputs by treating elasticity substitution as unknown parameter.[3] But the assumptions of homogeneity and additives remained in CES production function along with rigidity limitation on pattern of inputs substitution. These issues were later resolved by Uzawa (1962) and McFadden (1963).

The above advances in both types of production functions namely Cobb and Douglas and CES production function can be applied in economics analysis of organic farming. In agriculture economics for both organic and conventional farming empirical investigation of input output analysis; elasticity of substitution, price elasticity and returns to scale of factors are commonly estimated.

MODERN THEORIES OF PRODUCTION AND ORGANIC FARMING

In the modern era, various advances were made by economists such as Samuelson, Arrow, Shepherd, Hicks, Taylor and Jorgenson for the analysis of input–output relationship in production economics. Samuelsson (1947) developed the neoclassical hypothesis that how to allocate factors of production that minimize the cost of production and increase the productivity of inputs. This is popularly known as optimum allocation of resources. Another development was made that the cost curves were derived by using the production function.

The optimum allocation of organic inputs and minimum degradation of natural resources is the core issue of organic farming. So the above development which was made by Samuelson for other sectors is equally important for organic farming. For calculation of efficiency of factors of production, usually econometric processes were followed using some specific type of production functions. The simplest form

---

[3] Econometric studies based on the CES production function.

of production functions are the Cobb–Douglas function, CES and translog production function. The estimations of both the parameters, that is, cost and production function were used, but in the modern era cost function was more popular than production function. Arrow et al. (1961) derived cost curves using estimated parameter derived from the maximization of TC function. The neoclassical economists mostly used production function to derive TC curves and further used to find minima of TC curve. These technical issues were resolved in this era by the duality theory applied by Shephard (1953). In the duality theory, parameters are calculated using cost and profit functions along with share equations of inputs instead of using production function (Uzawa, 1964).

The invention of the duality theory using cost or profit function along with share equations eliminates the limitation of using production function only in economic modelling. The duality approach was primarily developed by Hotelling (1932). Later it was popularized by Samuelson (1954, 1960)[4] and Shephard (1953, 1970)[5] by empirical application of duality approach in agriculture and other sectors of economy.

In the process of development, the stochastic duality theoretical approach was developed by Taylor (1984) and its empirical application was also illustrated in the same paper. The two main features of the duality theory are: (a) to describe production function through dual representation, that is, either by cost or price function and (b) to construct explicit supply and demand functions as the derivative of cost or price function.[6] It has been found that the formulation of production theory using dual approach has the same implications of optimizing behaviour of producer that were in the study of Hicks in 1946 and Samuelson in 1983. So the duality theory helps to develop econometric methodology.

[4] Hotelling (1932) and Samuelson (1954) developed the dual formulation of production theory on the basis of the Legendre transformation. Employed by Jorgenson and Lau (1974a, 1974b) and Lau (1976, 1978a).

[5] Shephard utilizes distance functions to characterize the duality between cost and production functions. This approach is employed by Diewert (1974a, 1982), Hanoch (1978), McFadden (1978) and Uzawa (1964).

[6] Surveys of duality in the theory of production are presented by Diewert (1982) and Samuelson (1983).

The duality approach has been widely applied in the agriculture sector and other sectors of economy. It has been also applied to construct explicit function of relative prices by using demand and supply functions without imposing subjective constraints. It produces more efficient results in econometric modelling in production analysis (Jorgenson, 1986). Duality approach is now widely applied in organic farming and input–output analysis of other sectors of economy.

EMPIRICAL ENQUIRY ON PRODUCTION THEORIES IN AGRICULTURE AND ORGANIC FARMING

Abundant research has been made by Indian scholars using the Cobb–Douglas, CES and translog production functions in agriculture and organic farming. The application of production theories in organic farming is found after the 2000s because the movement of organic farming began after 2000 in India. The early works on application of production theories in agriculture and organic farming were by M. Alagh (2004), Srinivas and Ramanathan (2005), Mruthyunjaya, Rajashekharappa, Pandey, Ramanarao and Narayan (2005), Anupama, Singh, and Kumar (2005), Kamat, Tupe and Kamat (2012), Basavaraja, Mahajanashetti and Sivanagaraju (2008), Prajneshu (2008), Thakare, Shende and Shinde (2012) and Manjunath, Swamy, Jamkhandi and Nadoni (2013). The simplified form of the Cobb-Douglas production was used by the choosing only one dependent (X) and one independent variable (Y), shown as in equation (2.2) as follows:

$$Y = \alpha X^{\beta}, \qquad (2.2)$$

where α is parameter of scale and β is a measure of curvature. The parameters are estimated by using usual procedure. The function becomes linear when we take logarithmic of equation (2.2) and add the expected error term (ε) written as;

$$\ln(Y) = \ln(\alpha) + \beta \ln(X) + \varepsilon, \qquad (2.3)$$

The linear equation (2.2) can be written in the form of regression fitted equation, where the error term is shown by ε. The method of least square was used to calculate the coefficient of determination (R^2) to measure the goodness of fit of the model.

Alagh (2004) studied the responsiveness of price elasticities with respect to aggregate supply of agriculture food during 1950–1951 to 1996–1997 in India. He used aggregate supply response function to calculate price elasticity of output. He also pointed out that earlier scholars' result that agriculture supply function is not responsive to price. His result showed that after the 1980s a weak relationship was found between acreage response and terms of trade and that acreage response to price is market determined in the non-food sector. At the same time, it is highly significant in the model. So the agriculture response to price is stimuli at aggregate level.

Srinivas and Ramanathan (2005) conducted a research on the yield of elephant foot yam in three states, that is, Andhra Pradesh, Kerala and Tamil Nadu in 2002–2003. Their research was based on primary field survey results of 90 farmers collected from five main elephant foot yam cultivating districts. The found that the total cultivation cost was the lowest at ₹93,450 per hectare and varied from district to district in Andhra Pradesh. In Kerala it was the highest at ₹173,150 and in Tamil Nadu it was ₹168,032. The cost benefit ratio were 1.4, 1.5 and 1.4 respectively in the states.

Mruthyunjaya et al. (2005) conducted a field survey study in 14 major states of India for the years 2002–2003 and 2003–2004. A total 1,560 farmers are surveyed from these 14 states on four major edible oilseeds in India. These oilseeds are soybean, groundnut, rapeseed and mustard, and sunflower. They calculated technical inefficiencies in oilseed production to be 0.25–0.33 at average level and more at farm level. The collective technical inefficiency was found to be 0.5–0.66. In factors of production, the marginal returns to water is the highest on oilseeds. They suggested to increase the profitability on oilseed cultivation, it must be grown on irrigated land.

Anupama et al. (2005) conducted a research based on the Indian Agricultural Research Institute's (IARI) project for the year of 2000–2001. They calculated technological efficacy of maize crop for the state of Madhya Pradesh on 300 farms of three major maize cultivating districts: Shahdol, Mandsaur and Chhindwara. They used stochastic production frontier to calculate parameter, technical efficiency and adoption index to measure adoption of technology by the farmers. Their finding shows that the technology acceptance among the farmers was low. The cost comparison among tradition, composite and hybrid showed that the cost of cultivation increases with use of superior technology.

Kamat et al. (2007) have examined the determinants of agricultural gross domestic product (AGDP) for the period 1970–1971 to 2002–2003 of India. They applied the OLS regression estimation method using the Cobb–Douglas type production function and found that the agriculture sector recorded decreasing returns to scale in the 1970s, 1980s and 1990s in estimated type II model and constant returns to scale in in type I model during the 1970s and 2000s. They also found that after 1991 the input availability declined.

Prajneshu (2008) conducted study on time series data of Punjab for 29 years from 1971 to 2000. They tried to establish a relationship between inputs and output of wheat yields and tested to check goodness of fit of data using the Cobb–Douglas type production function. They calculated mean square error (MSE) value using the Cobb–Douglas type production function for one variable, which was 15.1 that showed goodness of fit in data. In case of two variables, the MSE value was 3.8, which was relatively low, showing goodness of fit in data.

Basavaraja et al. (2008) conducted a study based on a sample of 480 farmers in four major paddy growing districts of Andhra Pradesh during the kharif season in 2005–2006. They have calculated cultivation cost, revenue or income earn from sale and profit for both traditional and SRI methods paddy cultivation. They also applied the Cobb–Douglas type production function in both cultivation methods. Their finding showed that yield under SRI was 8.52 tons per hectare and 6.07 tons per hectare in traditional method. At the same time the cost of cultivation was recorded higher in SRI method than traditional method.

Thakare et al. (2012) used pool data for the period from 1999–2000 to 2008–2009 of cotton crops in Vidarbha region of Maharashtra. They tried to estimate response to change in demand and supply with respect to change in output and inputs price. They applied the normalized Cobb–Douglas type profit function to estimate inputs demand and supply of output. The inputs demand equations were derived from the profit function and the own inputs price elasticity of respective inputs have get expected negative sign as per theoretical formulation. The result showed that 1 per cent increase in own price of input, prices of other inputs remaining constant, the demand of fertilizer will decline at 2.1 per cent, demand of seed will decline at 1.7 per cent, and demand of labour and bullock both will decline and 1.3 per cent. The own price elasticity of all inputs is recorded greater than one.

Manjunath et al. (2013) conducted a field survey on a sample of 100 farmers in five villages of Hirekerur taluka of Haveri District of Karnataka State. They tried to investigate efficient use of input resource in both BT and non-BT cotton by using the Cobb–Douglas type production function. Their finding showed that seed and labour were the most significant inputs for increasing the gross returns of both types of crop cultivation. The regression coefficient for pesticide in both types of crops had a negative sign, meaning that if you increase the use of these inputs, gross returns will fall.

CONCLUSION

The economic analysis of producer behaviour is studied in different phases, namely, classical, neoclassical and modern. All these economic theories have been primarily applied and tested in the agriculture sector and now are being applied in organic farming. Each group of thought has applied different types of production function approaches to maximize producers' profit with respect to the given resources. The aim of all economic theories is the same but the method they used varied in each group. The most sophisticated analysis on production economics by using econometric technique is found in modem economic thought, which has been tested empirically across counties in agriculture and organic farming. Many empirical studies on agriculture and organic farming have been carried out in India.

REFERENCES

Alagh, M. (2004). Aggregate agricultural supply function in India. *Economic and Political Weekly, 39*(2), 202–206.

Anupama, J., Singh, R. P., & Kumar, R. (2005). Technical efficiency in maize production in Madhya Pradesh: Estimation and implications. *Agricultural Economics Research Review, 18*(2), 305–315.

Arrow, K. J., Chenery, H. B., Minhas, B. S., & Solow, R. M. (1961). Capital–labor substitution and economic efficiency. *The Review of Economics and Statistics, 43*(3), 225–250.

Basavaraja, H., Mahajanashetti, S. B., & Sivanagaraju, P. (2008). Technological change in paddy production: A comparative analysis of traditional and SRI method of cultivation. *Indian Journal of Agriculture Economics, 63*(4), 629–640.

Brown, M. (1966). *On the theory and measurement of technological change.* New York, NY: Cambridge University Press.

Cassels, J. M. (1936). On the law of variable proportions. In F.W. Taussig (Ed.), *Explorations in economics: Notes and essays contributed in honor of F.W. Taussig* (pp. 223–236). New York, NY: McGraw-Hill.

Cobb, C. W., & Douglas, P. H. (1928). A theory of production. *The American Economic Review, 18*(1), 139–165.

Danø, S. (1966). Diminishing returns and the cost function: A reconsideration. *Weltwirtschaftliches Archiv,* 97–115.

Dhrymes, P. J. (1965). Some extensions and tests for the CES class of production functions. *The Review of Economics and Statistics, 47,* 357–366.

Diewert, W. Erwin. (1974a). Applications of duality theory. In *Frontiers of quantitative economics,* ed. Michael Intriligator and David A. Kendrick. Vol. 2. Amsterdam: North-Holland.

Diewert, W. Erwin. (1982). Duality approaches to microeconomic theory. In *Handbook of Mathematical Economics,* ed. Kenneth J. Arrow and Michael D. Intriligator. Vol. 2. Amsterdam:North-Holland.

Douglas, P. H. (1948). Are there laws of production? *American Economic Review, 38*(1), 1–41.

Douglas, P. H. (1967). Comments on the Cobb-Douglas production function. In M. Brown (Ed.), *The theory and empirical analysis of production* (pp. 15–22). NBER. Retrieved from https://www.nber.org/chapters/c1474

Douglas, P. H. (1976). The Cobb–Douglas production function once again: Its history, its testing, and some empirical values. *Journal of Political Economy 84*(5), 903–916.

Felipe, J., & Adams, F. G. (2005). 'A theory of production' the estimation of the Cobb–Douglas function: A retrospective view. *Eastern Economic Journal, 31*(3), 427–445.

Fisher, F. M. (1971). Aggregate production functions and the explanation of wages: A simulation experiment. *The Review of Economics and Statistics, 53*(4), 305–325.

Hanoch, Giora. (1978). Symmetric duality and polar production functions. In *Production economics: A dual approach to theory and applications,* ed. Melvyn Fuss and Daniel McFadden. Vol. 1. Amsterdam: North-Holland.

Heady, E. O., & Dillon, J. L. (1961). *Agricultural production functions.* Ames: The Iowa State University Press.

Hicks, J. R. (1946). *Value and capital.* London: Oxford University Press.

Hotelling, H. (1932). Edgeworth's taxation paradox and the nature of demand and supply functions. *The Journal of Political Economy, 40*(5), 577–616.

Jorgenson, D. W., and Lawrence J. L. (1974a). Duality and differentiability in production. *Journal of Economic Theory 9:* 23–42.

Jorgenson, D. W., and Lawrence J. L. (1974b). The duality of technology and economic behavior. *Review of Economic Studies 41:* 181–200.

Jorgenson, D. W. (1986). Econometric methods for modeling producer behavior. In Z Griliches and MD Intriligator (Ed.), *Handbook of econometrics* (Vol. 3, pp. 1841–1915). Amsterdam: North-Holland Publishing.

Kamat, M. S., Tupe, & Kamat, M. (2007). Indian Agriculture in the New Economic Regime, 1971–2003: Empirics based on the Cobb Douglas Production Function, MPRA paper No 6150.

Kamat, M. S., Tupe, S. N., & Kamat, M. M. (2012). Indian agriculture in the new economic regime, 1971–2003: Empirics based on the Cobb Douglas production function (MPRA Paper No. 6150). Indian Institute of Technology Bombay, Goa University. Retrieved from https://mpra.ub.uni-muenchen.de/6150/1/MPRA_paper_6150.pdf

Krauthamer, S. (1963). Some ambiguities in the concept of fixed cost. *Economic Inquiry,* 2(1), 38–41.

Larson, B. (1991). A dilemma in the theory of short-run production and cost. *Southern Economic Journal,* 58(2), 465–474.

Lawrence, J. L. (1976). A characterization of the normalized restricted profit function. *Journal of Economic Theory 12:* 131–163.

Lawrence, J. L. (1978a). Applications of profit functions. In *Production economics: A dual approach to theory and applications,* ed. Melvyn Fuss and Daniel McFadden. Vol. 1. Amsterdam: North-Holland

Manjunath, K., Swamy, P. D., Jamkhandi, B. R., & Nadoni, N. N. (2013). Resource use efficiency of bt cotton and non-bt cotton in Haveri District of Karnataka. *International Journal of Agriculture and Food Science Technology,* 4(3), 253–258.

Maxwell, W. D. (1965). Short-run returns to scale and the production of services. *Southern Economic Journal,* 31(1), 1–14.

McFadden, D. L. (1963). Further result on C.E.S. production function. *The Review of Economic Studies, 30,* 73–83.

McFadden, Daniel. (1978). Cost, revenue, and profit functions. In *Production economics: A dual approach to theory and applications,* ed. Melvyn Fuss and Daniel McFadden. Vol. 1. Amsterdam: North-Holland

Mruthyunjaya, S. K., Rajashekharappa, M. T., Pandey, L. M., Ramanarao, S. V., & Narayan, P. (2005). Efficiency in Indian edible oilseed sector: Analysis and implications. *Agricultural Economics Research Review, 18*(2), 153–166.

Prajneshu, G. (2008). Fitting of Cobb-Douglas production functions: Revisited. *Agricultural Economics Research Review, 21*(2), 289–292.

Samuelson, P. A. (1947). *Foundations of economic analysis.* Cambridge, MA: Harvard University Press.

Samuelson, P. A. (1953–4). Price of Factors and Goods in General Equilibrium, *Review of Economic Studies, 21,* 1–20.

Samuelson, P. A. (1960). Structure of a minimum equilibrium system. In R. W. Pfouts (Ed.), *Essays in economics and econometrics: A volume in honor of Harold Hotelling,* (pp. 1–33). Chapel Hill, NC: University of North Carolina Press.

Samuelson, P. A. (1979). Paul Douglas's measurement of production functions and marginal productivities. *Journal of Political Economy*, 87(5), 923–939.

Samuelson, Paul A. (1983). *Foundations of economic analysis*. Cambridge, Mass.: Harvard University Press.

Sandelin, B. (1976). On the origin of the Cobb-Douglas production function. *Economy and History*, 19(2), 117–123.

Schultz, H. (1929). Marginal productivity and the general pricing process. *Journal of political Economy*, 37(5), 505–551.

Shephard, R. W. (1953). *Cost and production functions*. Princeton, NJ.: Princeton University Press.

Shephard, R. W. (1970). *Theory of cost and production functions*. Princeton, NJ.: Princeton University Press.

Srinivas, T., & Ramanathan, S. (2005). A study on economic analysis of elephant foot yam production in India. *Agricultural Economics Research Review*, 18(2), 241–252.

Stigler, G. (1939). Production and distribution in the short run. *Journal of Political Economy*, 47(3), 305–327

Taylor, C. R. (1984). Stochastic dynamic duality: Theory and empirical applicability. *American Journal of Agricultural Economics*, 66(3), 351–357.

Thakare, S. S., Shende, N. V., & Shinde, K. J. (2012). Mathematical modeling for demand and supply estimation for cotton in Maharashtra. *International Journal of Scientific and Research Publications*, 2(3), 1–5.

Tinbergen, J., Jorgenson D. W., & Waelbroeck J. (1978). Contribution to economic analysis. In Fuss M & Mafadeen D. (Eds.), *Production economics: A dual approach to theory and application* (Vol. 2). New York, NY: Oxford Publication.

Uzawa, H. (1962). Production functions with constant elasticities of substitution. *The Review of Economic Studies*, 29(4), 291–299.

Uzawa, H. (1964). Duality principles in the theory of cost and production. *International Economic Review*, 5(2), 216–220.

Viner, J. (1931). Cost Curves and Supply Curves. *Zeitschrift für Nationalökonomie (Journal of Economics)*, 3(1), 23.

Von Liebig, J. F. (1855). *Principles of agricultural chemistry: With special reference to the late researches made in England*. London: Walton & Maberly.

Walras, L. (1876). Un nuovo ramo della matematica. MATHEMATICAL ECONOMICS, 189.

Walras, L. (1900). Preface to the fourth edition. In *Elements of pure economics or the theory of social wealth* (p. 4).

Walras, L. (1900). *Elements of Pure economics or theory of social wealth*, fourth edition, UK, London (p. 4)

Walters, A. A. (1963). Production and cost functions: An econometric survey. *Econometrica*, 31(1), 1–66.

Farm Household Theories and Organic Farming

INTRODUCTION

This chapter presents the advances in household production theories and application in organic farming. Household production economic analysis has seen a 'knowledge gap' between observed choices and predicted choices of a farm household. The household production structure advocates that the cause of market failures and the ex-ante abilities provides explanations to some degree to manage risk. Thus the theoretical advances of 'behavioural economics'[1] have shaped a new research area which was missing in the neoclassical framework (Mendola, 2007). The neoclassical framework explains the risk–return trade-off faced by farm households in farming and emphasized the basic reasons behind it. The trade-off between risk and return faced by organic producers is common across countries due to fall in output. This explanation leads to the advances in modern behavioural approach to study the 'real people in real environments' (Roumasset, 2004). The farm household economic theories are classified into three major categories: The profit maximizing approach, utility maximizing approach and risk-averse peasant approach. All these household theories are widely used in organic farm household decision behaviour analysis.

These alternative economic theories of household behaviour illustrate the production and consumption choices of farm households. Each

---

[1] Behavioural economics: joint investigation of psychology and economics into the behaviour of a household.

theory assumes that a peasant[2] farm household's objective function is to maximize production under given constraints (Ellis, 1993) as organic producers aim to maximize their production under given organic inputs and organic policy constraints. These economic theories were based on various assumptions considering the larger agriculture economy within which peasant farm production takes place. Although all theories do not have common assumptions, they all follow the same explanation procedure of farm household behaviour. The profit maximizing theory of farm household has been criticized that it neglects the consumption aspect of the peasant farm household decision-making processes. However, the neoclassical household models includes both the consumption and production goals of farm household; as a result of which this approach became more popular. To counteract these models, other new liberal economists have formulated the risk aversion theory, which is based on the objective function of the farm households, that is, survival by avoiding risk. This is applicable to organic farming also (Mendola, 2007).

PROFIT MAXIMIZING THEORY AND ORGANIC FARMING

The theoretical contributions of Becker (1965) and Nakajima (1969) in the 1970s opened a door to further enquiry of household behaviour in economics. The theory of household production became popular as it considered farm household as the basic unit of food production and consumption. Here, the decision of organic farming is made by the farmer mainly due to self-consumption and environmental concerns. Empirical applications of farm household economics theory were established during the 1980s with the work of Schultz's (1964), Barnum and Squire (1979), and Singh, Squire and Strauss (1986). Mostly all applications of the farm household production model are based on a simple conceptual framework of household, product market, firm and labour market. The aim of the farm household is to maximize profit at given inputs, budget and production function technology constraints.

---

[2] Peasant farm households are those that are poor and do not have capital to adopt any new technology.

Schultz's (1964) has constructed a hypothesis that 'farm households in developing countries are poor but efficient.' Prior to Schultz's work on "Transforming traditional agriculture" (book) in 1964 it was assumed that peasant farm households are poor because they are backward and inefficient. Schultz has illustrated that a peasant farm household produces under profit maximization gaol and efficiency arises due to assumption of perfect competition in both input and output market. (i.e., farm producers supply at the given market price, price of inputs are equal to their marginal product, inefficient farm producers exist from farming business, and non-diminishing marginal utility of money). Schultz work resulted in a healthy debate on farm household behaviour and efficiency measurement among economists; as a result of which a lot of empirical works have been made to test it.

Many studies adopted the criteria of allocative efficiency and productive efficiency to test whether peasant farm households were efficient or not, and were they profit maximizers or not. Bliss and Stern (1982) found some contradictory results in case of Indian villages. They found conflicting evidence in efficiency and profit results. The main caution in profit maximizing approach is that it has both motivational and technical economic aspects. Generally, the empirical work on farm household production investigation was related with the efficiency measures of farm household, and less with way a farm household takes its decisions related to efficiency of the farm. In organic farm household's decision behaviour analysis profit maximizing theory is widely applied.

UTILITY MAXIMIZATION THEORY AND ORGANIC FARMING

The farm household profit maximization theory were criticized that profit maximization is not only single goal of farm producer but there is trade-off between profit maximization goals and other family goals. The goal of a farm household is utility maximization rather than profit maximization. The utility maximization approach have dual goals; maximization of family utility and enterprises rather than profit maximization only in case of organic farming and farm household anlysis. Chayanov (1966) explored the impact of family size and farm structure on a farm household's decision in the absence of a labour market. The absence of a labour market assumption has great influence on the economic analysis of farm production behaviour model. The absence of a labour market assumption in this approach, he used the value of labour time for work and hence the optimum use of labour is a subjective and varies across households as per their demographic and farm structure. He makes another strong assumption that the supply of land is unlimited. These assumptions are the limitation of the model (Chayanov, 1966).

The expansion of the Chayanovian model of farm household behaviour became popular during the 1960s. It explains farm household behaviour in simultaneous decision-making processes concerning production and consumption, based on the assumption of perfect markets structure. Becker's (1965) model of farm household behaviour typically included the notion of total household income. In this model household seen as a production unit that converts its own resources into final goods and services, which are available for sale and consumer purchase them to attain utility. Consequently, household maximize utility by consuming both home-produced goods and market-purchased goods, along with leisure time that is available after work, subject to his/her income constraints.

Becker's model explains that if a market is available for all goods, then they are tradable, prices are exogenous and decisions of production are taken by the household independently via use of time. According to him, the first-best choice is that where the marginal rate of transformation (MRT) in production is equal to the marginal rate of substitution (MRS) for each pair of goods in consumption. In such a situation, the decisions are consider as recursive because time used in production and time spent on leisure are treated independently, use of family labour is directly linked with the market wage rate, and income is the established link between consumption and production (Singh et al., 1986).

The reliability of Becker's recursive decision suffered in the absence of a labour market, as assumed in Chayanovian household behaviour model. The decision of time allocation between work and leisure may not be recursive because it is taken by the family. In Becker's household model, there is no separation between production and consumption activities. The decision process of household becomes circular as consumption affects income and income affects consumption. Then it is clear that the validity of recursive model of household time allocation depends on the assumption that household is a price taker and imperfect markets for input and output, including labour and capital, are absent (Mendola, 2007). Another study conducted by Bardhan and Udry (1999) on farm household's behaviour in developing nations are facing many market imperfections that prevent first-best business and investments. The field investigation result of farm household behaviour is not as per the theoretical expectation.

The debate on theoretical advances of farm household behaviour with missing market crafts a research area for neoclassical economists, but the objective of household remains unchanged. Utility maximization is the prime objective for all types of consumption goods and services, be it either home-produced or purchased along with leisure. The decision of organic farming is arising due to motivation of family members for self-consumption and natural resource conservation. Here, only total income is not the driver between consumption and production; there may be many other constraints, such as, market failure and missing market. Simultaneously, the task of empirical economics has shifted towards probing the market inefficiency and its impact on household production choices that were known as second-best (de Janvry, Fafchamps, & Sadoulet, 1991; Mendola, 2007).

THE RISK AVERSION THEORY AND ORGANIC FARMING

The above discussed theories are not able to explain a farm household's behaviour and have some serious shortcomings in explanation of farm household economies. Utility maximization theory is like profit maximizing theory and both have ignored the effect of uncertainty and risk on farm household's decision behaviour which are involved in any decision making like organic farming and other household activities. Both approaches are static in nature and assume certain equivalent constraints without considering risk. The empirical validation of farm household models focused on analytical tractability and availability of reliable data sources, simplified objective function of household and budget constraints. (Taylor & Adelman, 2003). The above approaches are criticized based on their ignorance of uncertainty and risk aversion which played a vital role in farm household cultivation decisions (Mendola, 2007) and choice of organic farming.

Farm households produce organic or non-organic crops under a high level of uncertainty because of natural hazards, namely, monsoon uncertainty, uneven rainfall, pest, climate change and plant diseases, fluctuations in market demand and uncertainty related to control of resources, for example, land holding and state's intervention through agriculture policy and war. All these uncertainties are causes of risk for a farm household, and it has to be cautious in its decision-making (Ellis, 1992; Walker & Jodha, 1986). So generally it assumed that a farm household avoids risk in its decision. Lipton (1968) has made disparagement with profit function approach. In his illustration of household behaviour, he incorporated risk and uncertainty in profit maximizing model. According to him, the small and marginal farmers are risk avoiders and try to fulfil basic family needs from production activities. There is less change that higher income level of household take risky decisions. (Lipton & Longhurst, 1989; Mendola, 2007).

Farm households' risk aversion theory can be conceptualized in two ways: first is the expected utility approach and second is the disaster avoidance approach. The expected utility approach deals with the decision-making of farm households under available risky alternatives at stated preferences of outcomes and probability. The production decision related to crop choice depends on the probability distributions with different outcomes under certainty and risk. This expected utility approach is seen in choice of organic farming decision by farm households. The neoclassical expected utility approach was based on the personal preference of decision-makers among the expected outcomes and occurrence of respective probabilities. Thus, the utility function of outcomes is concave to the origin, reflecting risk aversion and uncertainty. The expected utility also incorporates personal subjective desire for risk to explain production behaviour.

The standard expected utility theory approach is normative in nature and is based on various axioms, which are essential for the validation of von Neumann–Morgenstern's expected utility model and implicit utility maximization hypothesis (Mas-Colell, Whinston, & Green, 1995). Both consumption and production behaviour of a farm household and its revealed preference toward risk reflect its utility function. They have assumed the ceteris-paribus assumption; a risk avoider household prefers a smooth flow of consumption rather than a fluctuating one. In developing economies, where capital markets are not developed, financial institutions entail a low risk portfolio choice of producers (Morduch, 1994).

The validity of expected utility theory has been challenged by many empirical studies on peasant and organic farmer's behaviour. Extensive literature is available on systematic abuse of the von Neumann–Morgenstern expected utility approach and more recently behavioural and experimental inquiry has criticized its validity (Kahneman, Slovic, & Tversky, 1982). However, there is a different concern between experimental testing and real-life choices, namely, crop plantation and crop insurance. Roumasset (1976) has made early criticism on expected utility theory based on his own result of farmer's decision-making in Southeast Asia. His claim was on the method of risk aversion measurement which is only defined by the utility function independently. The absence of decision costs is another limitation. Furthermore, expected utility model can be described as a 'full optimality model' in which an individual makes the best choice under stated constraints. But this approach is unable to specify the decision-making process under which the decisions are made. Thus, this approach ignores the role of cost of production in the decision-making process. So the complexity of risks and uncertainty faced by farm producer in decision-making can help in further development of allocative choice models. On the other hand the disaster avoidance approach used for farm household decision behaviour does not have the ability to compute expected returns of available different alternative complex probability distribution outcomes.

The debate on normative or descriptive nature of expected utility theory is over and offer scope for farm household production theories.

The implicit assumption of farm decision is utility maximization, the expected utility theory turns to descriptive and probabilistic illustration (Roumasset, 1976). As a result, the optimum utilization approach has emerged as a weak model of producer decision behaviour, especially for small and marginal farm producer in developing economies. In case of organic farming decision behaviour, marginal and small farmer found the same. Furthermore, some studies were based on individual act of behavioural rules; where producers choose some limited objectives from their experience by a finite process described as 'rules of thumb' (Dasgupta, 1993).

The optimal allocation approach is also not free from criticism. It introduced the idea of low income farm producer living in an uncertain environment. This approach assumes that farm household chooses the least risky income stratum among the available alternatives. They first prefer safety and then choose the best alternatives via expected income or utility returns. Thus, the above argument invalidates that the expected utility and its assumption of continuity have not guaranteed to farm household to enter in the gamble of organic or farming profession. The expected utility model is based on a rule of thumb, which is known as safety first model of choice with risk and uncertainty. The producer first ensures his or her survival and wants to avoid the risk of income loss below a subsistence level. Hence, risk is considered as stochastic variable, that is, income has lesser value than some disaster level. This criterion presents a trade-off between choice of risky income stratum and available other low-risk alternatives. There are two optional solutions in the safety first model; either the producer minimizes his or her objective with having probability of disaster or maximizes return at stated constraint with probability of disaster (Ellis, 1992).

The disaster-avoiding behaviour of a present producer has a vertical section in the utility function and convexity around the threshold income level in expected utility model (Dasgupta, 1993). The vertical section shows amount of disutility associated with the loss of another unit of money (Masson, 1974). The expected utility concept has some exciting speculation in decision-making. In this approach if a producer considers a disaster serious enough, then he/she may be willing to invest a large share of his portfolio income in a project as the probability of return of project increases. In the same way, the safety first approach can be also used in lexicographic preferences, but ordering does not show a real-valued utility function, both discrete and continuous (Mas-Colell et al., 1995). So the individual behaviour based on expected utility at very low-level income has no conformity with stressful conditions. The disaster avoidance approach is useful in understanding individual choices under risk and uncertainty (Dasgupta, 1993).

Thus, the safety first approach included some specific variables in decision-making, that is, threshold income levels chosen via the expected utility. This approach does not follow the rules of 'pure behavioural approach'. But the outcomes drown from this approach are strong enough both in optimal and behavioural choices. This approach appears as a better descriptive tool for low-income farm producer when he/she is making risky choices. Although these two approaches are necessarily used in different procedures in practice, it depends on the choice and initial conditions. In broader perspective, the utility maximization theory is unable to emphasize problems such as extreme poverty, food insecurity and life of peasant farmers in most parts of the world, but the safety first theory explicitly captures these aspects of peasant behaviour in rural economies. In case of organic farming, both approaches have limited application.

EMPIRICAL STUDIES OF FARM HOUSEHOLD BEHAVIOUR AND ORGANIC FARMING

Binswanger and McIntire (1987) tried to explore the farm household and customary features of production in land-abundant tropics, simple farming technology and transportation cost. They have found that the population density and value of land (price of land) is positively related. Which further draws several inferences, such as, no productive asset is present with high collateral value, hence credit supply is principally limited; the crop insurance and storage facilities along with the growth of assets is essential to covariant risks; the semi-arid low-population density areas have cost advantage but extensive grazing is inefficient livestock management; abundance of land value is associated with unavailability of institutional credit which prevents the appearance of landless labour, vertical extension and client links are the cause of covariant risks.

Kumbhakar (1994) estimated the allocative and technical inefficiency by applying a profit function with endogenous and exogenous inputs. Using the translog production function he estimated efficiency of 227 sample farms in West Bengal. The maximum likelihood method was applied in the estimation of production function parameters. The result of profit maximization shows that the average technical efficiency was 75.46 per cent, while the technical efficiency in the best farm was recorded to be 85.87 per cent. On the other hand, in the concerned of allocative efficiency, mostly all farms recorded underuse of endogenous inputs, that is, fertilizer, manure, bullock labour and human labour.

Holloway and Ilbery (1996) elaborated the farmers' attitudes towards environmental changes, in case of global warming: how farmers made adjustment to change crop and farm management. He treated farmer as the farm manager and studied crop adoption behaviour between two crops navy beans and vining peas, with respect to change in temperature due to climate change. They have made forecast that global warming have both positive and negative impact on output of crops and farmers can reduce risk of negative impact of output by adopting climate change strategy and new varieties crops.

Austin et al. (1998) analysed the farmer's decision-making process with the help of mathematical models to predict and explain behaviour, using EEDMF survey data of 252 samples of farmers in east Scotland. He has applied various model of linear programming to explain the strategic aspect of farmer decision making behaviour and found that in each model different factors are significant. In case of computer access, use of technology plays a role in decision-making. In environmental behaviour, structure of farm size influences the farmer's decision to use machinery and chemicals.

Willock et al. (1999) studied farmers' behaviour using a multidisciplinary approach including psychology, economics, management and mathematics. They tested 1,000 farmers, using multivariate regression models for organic farming. They found that two explanatory variables scale of production and environmental farming policy influence the adaptation. Stress is another variable which was included in the variable scale of production. All three sets of scales, Edinburgh Farming Attitudes Scale (EFAS), Edinburgh Farming Objectives Scale (EFOS) and Edinburgh Farming Implementation Scale (EFIS) are related to farming business and personal characteristics of farming.

Chavas, Petrie and Roth (2005) have estimated economic efficiency of 115 surveyed farm households of three semi-urban villages in the periphery of capital city of Banjul, Gambia. They found that technical and locative efficiency are fairly high among the surveyed farm households generated through Tobit model.

Komicha (2007) has examined farm household economic behaviour using variables such as, saving, credit, input and output under imperfect market conditions. The result was based on 240 farm household survey data obtained from two districts of Southeastern Ethiopia during September to January in 2004–2005. The farm producer's saving behaviour and its explanatory variable show that nearly 37 per cent of the farm income is saved either in financial or physical assets form. His finding shows that informal credit source is dominant source of credit to households even though the interest rate was higher. The efficiency differences recorded 12 per cent between credit-unconstrained and credit-constrained farm households' ceteris paribus. Farm households have demanded credit for production and consumption purposes, but institutions prefer only precaution purpose.

Cole, Giné and Vickery (2017) studied Indian farm households based on a field survey and their insurance coverage effects on production decisions. They have found that crop insurance has significant impacts on cash crops choice by the educated farmers, who prefers high returns. Their results support the argument that financial inclusion can mitigate the real effects of non-insured production risk. They have also addressed the issue of low insurance adoption, instalment payment of insurance crops protect payout's commencing claims by family members.

CONCLUSION

In economic literature, different types of economic theories are available on household behaviour namely; profit maximizing theory, utility maximization theory and risk aversion theory. All these theories have different goals of organic and non-organic farm households and try to achieve them at given resources or budget constraints. Economists are applying functional relationship between satisfaction/utility and resources. This function relationship is estimated by using the objective function and related budget constraints. The empirical enquiry related to all these theories are presented above which are conducted by researchers in different countries for both organic and non-organic farming.

REFERENCES

Austin, E. J., Willock, J., Deary, I. J., Gibson, G. J., Dent, J. B., Edwards-Jones, G., ... Sutherland, A. (1998). Empirical models of farmer behaviour using psychological, social and economic variables. Part II: Nonlinear and expert modelling. *Agricultural Systems*, 58(2), 225–241.

Bardhan, P., & Udry, C. (1999). *Development microeconomics*. Oxford: Oxford University Press.

Barnum, H. N., & Squire, L. (1979). An econometric application of the theory of the farm-household. *Journal of Development Economics*, 6(1), 79–102.

Becker, G. S. (1965). A theory of the allocation of time. *The Economic Journal*, 75(299), 493–517.

Binswanger, H. P., & McIntire, J. (1987). Behavioral and material determinants of production relations in land-abundant tropical agriculture. *Economic Development and Cultural Change*, 36(1), 73–99.

Bliss, C. J., & Stern, N. H. (1982). *Palanpur: The economy of an Indian village*. Delhi: Oxford University Press.

Chavas, J. P., Petrie, R., & Roth, M. (2005). Farm household production efficiency: Evidence from the Gambia. *American Journal of Agricultural Economics*, 87(1), 160–179.

Chayanov, A. V. (1966). Peasant farm organization. In D. Thorner, B. Kerblay & R. E. F. Smith (Eds.), *The theory of peasant economy* (pp. 229–278). Homewood, IL: Irwin.

Cole, S., Giné, X., & Vickery, J. (2017). How does risk management influence production decisions? Evidence from a field experiment. *The Review of Financial Studies*, 30(6), 1935–1970.

Dasgupta, P. (1993). *An inquiry into well-being and destitution.* Oxford: Clarendon Press.

de Janvry, A., Fafchamps, M., & Sadoulet, E. (1991). Peasant household behaviour with missing markets: Some paradoxes explained. *The Economic Journal, 101*(409), 1400–1417.

Ellis, F. (1992). *Peasant economics.* Cambridge: Cambridge University Press.

Ellis, F. (1993). *Peasant economics: Farm households in agrarian development* (Vol. 23). Cambridge: Cambridge University Press.

Holloway, L. E., & Ilbery, B. W. (1996). Farmers' attitudes towards environmental change, particularly global warming, and the adjustment of crop mix and farm management. *Applied Geography, 16*(2), 159–171.

Kahneman, D., Slovic, P., & Tversky, A. (Eds). (1982). *Judgment under uncertainty: Heuristics and biases.* New York, NY: Cambridge University Press.

Komicha, H. H. (2007). *Farm household economic behaviour in imperfect financial markets* (Doctoral Thesis). Swedish University of Agricultural Sciences, Uppsala (Vol. 2007, No. 78).

Kumbhakar, S. C. (1994). Efficiency estimation in a profit maximising model using flexible production function. *Agricultural Economics, 10*(2), 143–152.

Lipton, M. (1968). The theory of the optimizing peasant. *Journal of Development Studies, 4*(3), 327–351.

Lipton, M., & Longhurst, R. (1989). *New seeds and poor people.* London: Unwin Hyman.

Mas-Colell, A., Whinston, M. D., & Green, J. R. (1995). *Microeconomic theory* (Vol. 1). New York, NY: Oxford University Press.

Masson, R. T. (1974). Utility functions with jump discontinuities: some evidence and implications for peasant agriculture. *Economic Inquiry, 12*(4), 559–566.

Mendola, M. (2007). Farm households production theories: A review of 'institutional' and 'behavioural' responses. *Asian Development Review, 24*(1), 48–68.

Morduch, J. (1994). Poverty and vulnerability. *American Economic Review, 84*(2), 221–225.

Nakajima, C. (1969). Subsistence and commercial family farms: Some theoretical models of subjective equilibrium. In C. R. Wharton (Ed.), *Subsistence agriculture and economic development* (pp.165–185). Chicago, IL: Aldine Publishing.

Roumasset, J. (1976). *Rice and risk: Decision making among low-income farmers.* Amsterdam: North-Holland Publishing.

Roumasset, J. (2004). Rural institutions, agricultural development and pro-poor economic growth. *Asian Journal of Agriculture and Development, 1*(1), 56–75.

Schultz, T. W. (1964). *Transforming traditional agriculture.* Chicago, IL: University of Chicago Press.

Singh, I., Squire, L., & Strauss, J. (1986). *Agricultural household models: Extensions, applications, and policy.* Baltimore, MD: Johns Hopkins University Press for the World Bank.

Taylor, J. E., & Adelman, I. (2003). Agricultural household models: Genesis, evolution, and extensions. *Review of Economics of the Household, 1*(1–2), 33–58.

Walker, T., & Jodha, N. (1986). How small farmers adapt to risk. In P. Hazell, C. Pomareda & A. Valdez (Eds.), *Crop insurance for agricultural development.* Baltimore, MD: Johns Hopkins University Press.

Willock, J., Deary, I. J., Edwards-Jones, G., Gibson, G. J., McGregor, M. J., Sutherland, A., … & Grieve, R. (1999). The role of attitudes and objectives in farmer decision making: Business and environmentally-oriented behaviour in Scotland. *Journal of Agricultural Economics, 50*(2), 286–303.

Producer Behaviour, Resource Allocation and Organic Farming

INTRODUCTION

Producer behaviour is guided by various economic and non-economic factors in agriculture and organic farming. But for theoretical interpretation, economic variables, such as price, income, test preference, and prices of related organic and non-organic commodities, are considered to determine the market demand and supply of a specific product. Market demand encourages producer to produce and allocate the available resources in the production of that product. If the demand of organic products is increasing, the producer will shift from non-organic to organic farming. The producer choice of crop and cultivation method is motivated by profit, which can increase only either by reduction in cost of production or increasing production without increasing inputs. The econometric modelling of producer behaviour requires parametric forms of demand and supply function. These parameters represent the change in production pattern due to change in price, technology and scale. Econometric models of producer behaviour consider systems of demand and supply functions. All the dependent variables in these functions depend on the same set of independent variables. However, the variables and the parameters may enter the functions in a nonlinear manner. Efficient estimation of these parameters uses systems of nonlinear simultaneous equations (Jorgenson, 1986).

To know the allocation of resources in the available various alternative uses is necessary to understand the choices of household. The choices of a household are more or less influenced by elderly family members. Many economic studies have enquired the resource allocation between husband and wife at household level, which is another aspect household behaviour. Becker (1965) was the prominent economist who studied the household behaviour at family level and his theory of time allocation is a predominant work in new household economics. Here, the aim of this chapter is to present the works which are related to allocation of resources in agriculture and organic farming to attain maximum production. Agriculture is the early studied sector where the farmer has alternative choices related to crops and farming methods. The choice of farming method, organic or conventional, has greater degree of resource allocation than choice of crops.

THEORIES OF RESOURCE ALLOCATION AND ORGANIC FARMING

The 1975 Nobel Prize in economics was awarded to academician Leonid V. Kantorovich, USSR, and Professor Tjalling C. Koopmans, USA, for their contribution to the theory of optimum allocation resources. Their work was the beginning in this field and tackled the fundamental problem of all economic activities—how the available productive resources can be used for the maximum production of goods and services. This field answered questions such as what goods should be produced, what methods of production should be used, and how much of current production should be consumed and how much should be reserved to create new resources for future production and consumption. They formulated the relationship between productive inputs and output in new ways, and were able to achieve highly significant results. The theory of optimal allocation resources is widely used in agriculture and organic farming.

In his prior research, Professor Kantorovich had applied linear programming as an analytical technique to demonstrate how economic planning could be improved in a country. He made his first contributions in the field of economic research in 1939 when he wrote an essay on the meaning and significance of efficient use of resources in individual enterprises. In his book *The Best Use of Economic Resources*, he analysed similar efficiency conditions for an economy, and particularly demonstrated the connection between the allocation of resources and the price system, both at a certain point in time and in a growing economy.

Professor Koopmans, for his part, applied certain efficiency criteria to make important deductions concerning optimum price systems. In his primary work *Analysis of Production as an Efficient Combination of Activities*, so-called activity analysis, he developed new ways of interpreting the relationship between inputs and outputs of a production process which are used to clarify the declaration between efficiency in production and the existence of a system of calculation prices. This presents a new and interesting connection between the normative allocation theory and the general equilibrium theory.

The basic economic problems are more or less the same in all societies and places, regardless of whether it is a capitalist, socialist or mixed political organization. As the supply of productive resources is limited in all types of political system, a series of questions are raised concerning the optimal use of available resources and the fair distribution of income among citizens. The answer to such normative questions is more or less the same for each type of political system, but the problem becomes more serious when the further use of these resources in future is doubtful. Professor Kantorovich and Koopmans studied different types of political economics, Soviet Union and USA, but their choice of problems and methods was same. Both studied efficiency of production and their analysis is independent to each other, but they developed similar models of production. The efficiency study of organic and non-organic farming has applied this approach.

BECKER'S THEORY OF TIME ALLOCATION

A theory of the allocation of time by Gary Becker (1965) presented an analytical foundation for the study of household production and the allocation of time within the household. Becker's model of household formally engages in activities producing outputs such as food, children and housing (Z_i for commodity i) that bundle goods and time. The household consumes these commodities as direct objects of utility. The outputs of the activities were produced by distinct inputs. In the household model of production, he assumed to produce and consume a vector of commodities $Z = (Z_i)$, $i = 1, \ldots, I$. These commodities are associated with different levels of activities performed by the household (e.g., consumption of food, child-rearing, leisure activities, cooking of food). Utility is a function of these commodities:

$$U\,(Z_1, \ldots, Z_I), \tag{4.1}$$

where

$$Z_i = f^{(i)}\,(X_i, T_i), i = 1, \ldots, I, \tag{4.2}$$

where X_i is a vector of goods used to produce Z_i and T_i is time. The price of Z_i depends on the prices of its components. In the model it is assumed that each $f^{(i)}$ is homogeneous of degree one, then he has constructed a scale-invariant price index π_i for each commodity.

Each household faces two types of constraints, time and budget. Becker uses elementary algebra to explain household decision to choose constraints under the assumption that the household effectively faces one constraint. Under the assumption that T_i is scalar, and that the price of time is w across all uses, the maximum amount of income that the person can earn is full income $B = wT + V$ where $T = \Sigma T_i$ and V is the amount of unearned income accruing to the household. The Z_i includes all activities in which time can be used, including the consumption of leisure on the job and

$$\sum_{i=1}^{I} \pi_i Z_i = wT + V = B. \tag{4.3}$$

The household maximizes (4.1) subject to (4.2) and (4.3). The demands for inputs X_i, T_i are derived from the demands for Z_i. The responsiveness of the demands for different activities depends on response to changes in the prices of goods and time. Becker has presented a more general analysis where the marginal cost of time varies across activities.

In Becker's model of commodity, demand is an instance of Gorman's (1959) general separability analysis where U is weakly separable in the arguments producing the Z_i and the $f^{(i)}$ are homogeneous of degree one. Under homogeneous weak separability, consumer decision-making can be characterized by a two-stage budgeting process. Agents allocate budgets E_i to each commodity i, based on the price index π_i and in a second stage maximize each Z_i subject to these allocations determined from the first stage to determine X_i and T_i (Gorman, 1959; Strotz, 1957). Pollak and Wachter (1975) present a definitive analysis of the limitations of the Becker model when the assumption of homogeneous separability is relaxed and when joint production is considered.

GAME THEORY, RESOURCE ALLOCATION AND ORGANIC FARMING

Game theory model has been applied in every field organic farming, agriculture, industry and social choice where people have multiple or more than one choice. Under appropriate assumption about the state of information, these naive models can be subsumed under the more general theories of decision-making based on von Neumann and Morgenstern's theory of games (Luce & Raiffa, 1957; Shubik, 1952; Tintner, 1942; von Neumann & Morgenstern, 1947) and Wald's theory of statistical decision functions (Blackwell & Girshic, 1954; Chernoff & Moses, 1959; Wald, 1950). Savage's subjective probability approach (Anscombe, 1961; Koopmans, 1957; Savage, 1954; Schlaifer, 1959) to risky choice. These approaches aim to maximize the decision maker's utility; stressing 'rational' behaviour, they have been postulated as normative theories.

The earliest use of game theory was made in poker, war and business. Like other decision-makers, a farmer rarely has full information about rainfall, productivity and other related inputs of production. The farmer's decisions related to crop choice, faming method—organic or conventional—and yield are based on past decisions. Removing this gap between actual and possible achievement is the normative aspect of decision-making. These problems are studied under descriptive economics, which deals with how are decisions made. These problems are studied using a variety of theories of choice for individuals assuming with and without risk-taking constraints. Game theory has been applied to study a range of agricultural problems such as: (a) production decisions under free competition, (b) the development of vertical and horizontal integration, (c) production under climatic uncertainty, (d) decisions on whether or not to adopt a new production technique of organic farming, (e) trading or bargaining activities and (f) conflict within the firm between its household and business sectors (John, 1962). Here we discuss only some problems which are more relevant for the producer decision-making such as, production decisions under free competitions, production under climatic uncertainty and decision on whether adopt a new production technique of organic farming.

1. Decisions under free competition: Farmers operate under free competition; no farmer can influence the price he pays or receives in the market. As Shubik (1952) has noted that no individual could possibly evaluate all his alternatives (such as in terms of product type, quality and time of marketing) relative to the alternatives open to each of his many competitors considered individually. Not only would there be too many people involved, (i.e., an n-person game with n being very large) but there would usually be a very large set of alternative strategies available to the decision-maker himself. Recognizing the impossibility of full information, Dillon and Heady (1962) have argued explicitly and Baker (1960) implicitly that production decisions by free competitors should be treated as games against Nature.

Under the given impossibility of full information, they suggest a rational approach for an individual producer would be to consider his alternatives relative to an 'aggregate' opponent made up of all the other members of the freely competitive group taken in combination with other sources of influence such as the weather (Moglewer, 1962). Although both players would have many alternatives strategies and payoff. Accordingly, it is argued that the decision-maker should first stratify both his own and Nature's alternatives, and then amalgamate those within each stratum into a single broad possibility. With payoffs represented by the most likely or expected payoff under each broad alternative, the net result of these simplifications would be reduction of the payoff matrix to a workable size (Bruner, Goodnow, & Austin, 1956; Haring & Smith, 1959; Simon, 1959).

The missing full information about the price expectations, planned choices and general approach of his confreres to the decision problem, the decision-maker could not associate objective probabilities with the broad alternatives of his aggregative opponent. In consequence, the payoff matrix depicts a game against Nature which might be tackled either as a decision problems under uncertainty (DPUU), or if some states of nature are thought to be more plausible than others, then as a subjective risk problem. Thus, this game theoretic model of free competition has the merit of incorporating uncertainty as an endogenous factor. In contrast, the classical theory of free competition completely ignores the possibility of uncertainty (John, 1962).

2. Vertical and horizontal integration: Baker (1960) and Dillon (1960) have suggested that the phenomena of vertical integration and farm cooperatives might be analysed in terms of game theory. For example, integration might be viewed as the development of coalitions in an n-person game being played by farmers, processors, retailers and consumers. Certainly, given such a chain of entrepreneurs whose decisions interact, coalitions of one type or another would be expected. As Davis and Whinston (1962) have argued such coalitions might be motivated by the desire to internalize external economies, although the reduction of uncertainty and the desire for countervailing power are probably more potent forces. From a policy and researcher point of view, the game theory approach to integration leads directly to important questions, for example, how viable are the coalitions? What are their effects on those in them and on those outside them? Are there better ways of playing the game? To some extent these questions provided that optimal solutions to n-person games are unavailable. However, this may be cause of further studying integration within a game theoretic framework. Not only is integration, either horizontal or vertical, a feasible solution to the game but is also an actual one that can be watched through its formative stages. As well, there are any number of suggestive n-person concepts which might usefully be drawn to the attention of public executives concerned with integration (both inter and external) and allied problems (John, 1962).

3. Production under climatic uncertainty: The consideration of production decisions under climatic uncertainty has been the most popular application of game theory in agricultural economics. The most extensive study was made by Walker and Heady (1960). They applied DPUU criteria to real-world situations involving choice between crop varieties, kinds and amounts of fertilizer, crop enterprises, pasture mixtures and stocking rates. Typically, the situations were such that insufficient climatic records were available to yield reasonable objective probabilities for Nature's choices. However, this approach suffers from the difficulty—common to many games against Nature—that some of Nature's states will often be unknown. If so, any derived solutions may be quite misleading.

The principal contribution of Walker and Heady was not the mere casting of climatic uncertainty problems into a game theoretic framework. That had been done before by Schickele (1950), Swanson (1957, 1959) and Thompson (1956), all of whom also ignored the possibility of a subjective probability approach. Rather, Walker and Heady's contribution was to demonstrate that if farm advisers used the various decision criteria, they could give alternative recommendations suited to a wide range of farmer attitudes and goals—especially if the recommendations were couched in terms of the practical characteristics of the criteria. Thus, Walker and Heady suggested:

a. The Laplace criterion is pertinent if the farmer is financially free to follow choices which may lead to highest long-run profits
b. The Hurwicz pessimism–optimism approach is relevant for optimistic farmers who desire and can afford to gamble
c. The Wald maximin criterion is best for farmers who must consider short-run outcomes because of financial commitments
d. The Savage-Niehans minimax regret algorithm is appropriate for farmers who cannot completely ignore short-run outcomes, but can give some weight to long-run profit considerations or are in a sufficiently flexible position to allow for the possible opportunity cost of alternative strategies.

Somewhat similar suggestions have also been made by Throsby (1961) based on a game against Nature case study of a farmer's problem in trying to decide when he should sell his livestock in the face of drought possibilities. However, such suggestions are no more than attempts to translate into pragmatic terms the axioms underlying the various decision criteria. So far as they lack mathematical precision, these endeavours are likely to be incomplete if not incorrect.

4. Decisions of whether to adopt a new production technique: The decisions about whether to adopt a new production technique of organic farming (innovation) imply complete ignorance in literature. Dillon and Heady (1958) have examined innovation or adoption of new technique as a DPUU. They formulate the payoff matrix as showed in Table 4.1.

The decision-maker's alternatives are self-explanatory. For Nature, the only relevant states are the extreme ones of absolute success or absolute failure. The decision-maker faces two types of payoff under the adoption: success and failure. Non-adoption yields zero payoffs regardless of whether the new technique of production or innovation would have succeeded or failed. For convenience, the payoff matrix is normalized in terms of the payoff under adoption if Nature decrees absolute failure. Assuming that a degree of experimentation (partial adoption) a, $(0 < a < 1)$, results in a payoff of a times the adoption payoff, application of the DPUU criteria yields supposedly optimal strategies. Normatively, however, these solutions are only worthwhile to the extent that reasonable estimates can be made of the adoption payoff. By the nature of the case, this may not be possible. And even if based on correct payoffs, the solutions are only strictly relevant for a decision-maker with absolutely no prior information about the innovation's chances of success. Such complete ignorance is rare for new agricultural practices; some vague information on the chance of

Table 4.1 *Payoff Matrix for Decision-makers*

Decision-Maker's Alternatives	Nature's Alternative	
	Success	Failure
Adoption	Z	−1
Non-adoption	0	0

Source: John (1962).

varying degrees of success is usually available so that a better procedure would more often be the use of subjective probabilities with a 'wider' payoff matrix (John, 1962).

The DPUU solutions to the innovation problem follow the pattern expected from the nature of the criteria. The Wald criterion, based on the assumption that Nature will do her worst, always dictates non-adoption. The Laplace criterion, placing equal weight on success or failure, suggests non-adoption or adoption according as the possible gain is less or more than the absolute value of the possible loss. The minimax regret algorithm, concentrating on ex-post consideration of foregone gains, tends to emphasize mixed (i.e., experimentation) strategies. Lastly, the Hurwicz approach suggests either non-adoption, experimentation, or adoption, depending on the relative sizes of the pessimism index b and the payoff for successful adoption (John, 1962).

As suggested by Emery and Oeser (1958) and Ruttan (1960) such game theoretic results may be useful descriptively as an analytical base for the study of the adoption of new agricultural techniques. Thus, farmers who never innovate may be following some Wald-type 'expect the worst' approach based on wait-and-see attitudes; farmers who follow an experimentation or partial adoption approach may do so because of opportunity cost considerations; while perennial innovators might be characterized by a Hurwicz-type approach with a high degree of optimism. Although no data specifically oriented to these hypotheses is available, there is some tentative evidence that supports them. For instance, Fallding (1958) and Rogers (1958) have classified farmers in terms of their use of recommended farm practices along lines suggestive of any classification based on the DPUU criteria (John, 1962).

TECHNOLOGY ADAPTATION BEHAVIOUR AND ORGANIC FARMING

The decision of producer to choose output and method of production is not easy under risk and uncertainty. Static economic theory provides guidance for making decisions when consumers have complete knowledge. Farmer decision to adopt organic or conventional farming, involves the optimal allocation of resources and efficient use of inputs, etc. (Walker & Heady, 1960). Like other decision-makers, farm households rarely have complete information related to projection of crop, weather, monsoon, climate, rainfall and soil contents. In farmer's decision, uncertainty will always be present and his present planning is based on about past experiences. Minimizing the gap between actual and expected returns is decision-making (Dillon, 1962).

The problems of decision-making have been recognized several decades ago and have led to the theoretical development of producer choice in risk-bearing situation (Heady, 1952, Pt. 11; Knight & Greve, 1960; see also the pioneering risk analyses of Hart [1940], Hicks [1931] and Knight [1957]).

The traditional and most precise theories stressed on the application of a simple rule of thumb, which was based on the projection of some weighted average of past results. Schultz and Brownlee (1942), Heady (1952) and Darcovich (1958) have listed a number of rules, such as the information availability, naive decision models based on game theory (Luce & Raiffe, 1957; Newman & Morgenstern, 1955; Shubhik, 1959; Wagner 1958), the theory of statistical decision functions of Wald (Blackwell & Girshick, 1954; Chernoff & Moses, 1959; Wald, 1950), and Savage's subjective probability approach (Anscombe, 1961; Koopmans, 1957; Savage, 1954; Schlaifer, 1959) to risky choice.

The aim of all these approaches is to maximize utility of decision-makers by choosing the best option. Shackle searched the gap and proposed a descriptive theory, which emphasizes a psychological nature of decision-maker's degrees of belief and potential outcomes (Akermans, 1958; Shackle 1949, 1958, 1961). Simon's theory tried to emphasize mathematical procedures using psychological variables (Simon, 1957). In Shackle's descriptive theory, the decision-maker seeks to satisfy a minimum level of objective rather than maximization of utility. Later Arrow (1951, 1958, 1959) and others made critical reviews of these theories (Savage, 1954; Simon, 1959). Edwards (1954) gave expository survey on less-recent literature, while Bross (1953) and Hildreth (1957) have argued for generalized decision-making, which has several impacts in the agricultural production analysis.

Over the decades, economists have made several attempts to test the degree of variability using both normative and descriptive approaches. Both Simon's and Shackle's descriptive theories are based on psychological variables, but there is little evidence to show that they play any great role in farmer decisions (Swanson, 1951; Dillon, 1960). On the other hand, there is some evidence of using naive expectation models by farmers and benefits gained by using such rules of thumb (Darcovich, 1958; Kaldor, 1958; Papandreou et al., 1957).

Mukhopadhyay (1994) tried to analyse technological change and household well-being at household level using the process of intra-household resource allocation among the family members and activities. He pointed that labour force participation is the main household activity in the local market and work for his/her own household production and care activities. The empirical findings of the study suggested that farm household' decisions to adopt new technology mainly depend on fertility of land, access to irrigation and expected returns. The wage of female labour is positively related to the proportion of land under HYV, while wage rate of male labour is negatively related to HYV cultivated area.

Oladele (2005) explored the farm household discontinuation behaviour of innovation and its influencing factors based on the 120 households survey data collected from two states of Southwestern Nigeria during 2002 arable crop growing season. The estimation result of Tobit regression model shows that the input and market availability variable having positive impact not allowed producers to discontinuation.

Singh, Barman and Varshney (2011) studied 100 vegetable growers in Jabalpur district of Madhya Pradesh. They reported that most of the growers have high and medium level knowledge of superior technologies of cauliflower and tomato cultivation. But they also reported that growers had some problems such as lack of weed control methods, leaf curl in tomato, high cost of fertilizer and pesticide, lack of marketing, infected seeds and unavailability of labour, which were the major causes for non-adoption of superior technologies.

Karki et al. (2011) investigated the adaptation behaviour of organic tea producers and their influencing factors. They surveyed a total of 181 farmers, 81 were organic and 86 conventional growers in the Ilam and Panchthar districts of Nepal. They applied discriminant analysis to identify socio-economic variables which distinguished organic growers to conventional growers. Their result suggested that tea growers located near regional markets, older in age, trained, affiliated with institutions and the size of their farm are variables which influence positively to adopt organic practices.

Akudugu, Guo and Dadzie (2012) examined farm households' adoption behaviour of modern technology and its influencing factors. The study was based on field survey data of 300 farm households in Bawku West district of Ghana. The logit regression model was applied to estimate relationship between adoption and it is influencing factors. Their empirical result showed that farm size, access to credit, expected benefits from technology and extension of services are positively related to adoption of technology.

Rana, Parvathi and Waibel (2012) examined the factors that motivate to adopt pepper organic farming in Idukki district of Kerala in India. The data was collected 100 organic farmers and 100 conventional farmers, with a farm size below three hectares during February and March 2011 for the year 2010. For the analysis logit regression analysis was used and it was found that agricultural area, assets, access to credit, and the perception of the household head towards adopting organic farming.

CONCLUSION

Producer choice and optimum resource allocation is found in agriculture and organic farming which are studied under microeconomics. Kantorovich and Koopmans had jointly received Nobel Prize for theory of optimum resource allocation in 1975. The use of game theory in solving strategic problems of choosing best allocation of resources is also found in empirical economic literature in agriculture and organic farming. In example of agriculture economics, the choice of crop and farming method under climatic uncertainty are also studied which involve risk in production decisions. The adoption of new technology of organic farming in agriculture depends on various factors such as farmer's education, income, farm size, availability of irrigation facility and public policy. These variables which are mentioned above influences positively to adaptation of new technology.

REFERENCES

Akerman, J. (1958). Professor Shackle on economic methodology, *Kykh* 11: 341–61,

Akudugu, M. A., Guo, E., & Dadzie, S. K. (2012). Adoption of modern agricultural production technologies by farm households in Ghana: What factors influence their decisions? *Journal of Biology, Agriculture and Healthcare*, 2(3), 1–9.

Anscombe, F. J. (1961). Bayesian statistics. *American Statistician*, 15(1), 21–24.

Arrow, K. J. (1951). Alternative approaches to the theory of choice in risk-taking situations. *Econometrica*, 19(4), 404–437.

Arrow, K. J. (1958). Utilities, attitudes, choices. *Econometrica*, 26(1), 1–23.

Arrow, K. J. (1959). Functions of a theory of behaviour under uncertainty. *Metroeconomica*, 11, 12–20.

Baker, C. B. (1960). Decision Making and Financing Farm Assets. J. S. McLean Memorial Lecture, Ontario Agricultural College, Feb. (Mimeo.)

Becker, G. S. (1965). A theory of the allocation of time. *The Economic Journal*, 75(299), 493–517.

Blackwell, D., & Girshick, M. A. (1954). *Theory of games and statistical decision.* New York, NY: Wiley.

Bross, I. D. (1953). *Design for decision.* New York, NY: Macmillan.

Bruner, J, Goodnow J. J., & Austin, G. A. (1956). *A study of thinking.* New York, NY: Wiley. Chapters 4 and 5.

Chernoff, H., & Moses, L. E. (1959). *Elementary decision theory.* New York, NY: Wiley.

Darcovich, W. (1958). Evaluation of some naive expectations models for agricultural yields and price. In M. J. Bowman (Ed.), *Expectations, uncertainty and business behaviour* (pp. 199–202, Chapter 14). New York, NY: Social Science Research Council.

Davis, O. A., & Whinston, A. (1962). Externalities, welfare and the theory of games. *Journal of Political Economy*, 70, 241–262.

Dillon, J. L. (1960). Integration and game theory. *Journal of Farm Economics*, 42(2), 384.

Dillon, J. L. (1962). Applications of game theory in agricultural economics: Review and requiem. *Australian Journal of Agricultural and Resource Economics*, 6(2), 20–35.

Dillon, J. L., & Heady, E. O. (1958). Decision criteria for innovation. Australian Journal of Agricultural Economics, 2(2), 113–120.

Dillon, J. L. and Lloyd, A. G. (1962). Inventory analysis of drought reserves for Queensland graziers: some empirical analytics, *Australian Journal of Agricultural and Economics*, 6: 50–67.

Edwards, W. (1954). The theory of decision making. Psychological *Bulletin*, 51(4), 380–417.

Emery, F. E., & Oeser, O. A. (1958). *Information, decision, and action.* Melbourne: Melbourne University Press.

Fallding, H. (1958). *Precept and practice on north coast dairy farms* (Research Bulletin 2). Department of Agricultural Economics, University of Sydney.

Gorman, W. M. (1959). Separable utility and aggregation. *Econometrica: Journal of the Econometric Society*, 27(3), 469–481.

Haring, J. E., & Smith, G. C. (1959). Utility theory, decision theory and profit maximization. *The American Economic Review*, 59, 566–583.

Hart, A. G. (1940). Anticipations, uncertainty, and dynamic planning, *Studies in Bus. Admin, 11: 1–98.*

Heady, E. O. (1952). *Economics of agricultural production and resource use.* New York, NY: Prentice-Hall.

Hicks, J. R. (1931). The theory of uncertainty and profit, *Economica 11: 170–89.*

Hildreth, C. (1957). Problems of uncertainty in farm planning. *Journal of Farm Economics, 39*(5), 1430–1434.

Jorgenson, D. W. (1986). Econometric methods for modeling producer behavior. In Zvi Griliches & Michael Intriligator (Eds.), *Handbook of econometrics* (Vol. 3, pp. 1841–1915), Elsevier: North Holland.

Kaldor, D. R. and Heady, E. O. (1954). An Exploration of Expectations, Uncertainty, and Farm Plans in Southern Iowa Agriculture. *Iowa Agriculture and Home Economics Experiment Station Research Bulletin,* 408.

Karki, L., Schleenbecker, R., & Hamm, U. (2011). Factors influencing a conversion to organic farming in Nepalese tea farms. *Journal of Agriculture and Rural Development in the Tropics and Subtropics* (JARTS), 112(2), 113–123.

Knight, F. H. (1957). Institutional Economics: Discussion. *American Economic Review, 47:* 18–21.

Knight, D. A. and Greve, R. W. (1960). *Use and Interrelation of Marginal Analysis to other Analytical Processes by Farmers in Decision Making.* Kansas State Univ. Agric. Exp. Sta. Tech. Bull. 108.

Koopmans, T. C. (1957). *Three essays on the state of economic science.* New York, NY: McGrawHill.

Luce, R. D., & Raiffa, H. (1957). *Games and decisions.* New York, NY: Wiley.

Moglewer, S. (1962). A game theory model for agricultural crop selection. *Econometrica, 30*(2), 253–266.

Mukhopadhyay, S. K. (1994) Adapting Household Behavior to Agricultural Technology in West Bengal, India: Wage Labor, Fertility, and Child Schooling Determinants, *Economic Development and Cultural Change, Vol. 43, No. 1, pp. 91–115*

Oladele, O. I. (2005). A Tobit analysis of propensity to discontinue adoption of agricultural technology among farmers in Southwestern Nigeria. *Journal of Central European Agriculture, 6*(3), 249–254.

Papandreou, A. G. et al. (1957). Test of a stochastic theory of choice. *University of California Publications in Economics, 16*(1), 1–18.

Pollak, R. A., & Wachter, M. L. (1975). The relevance of the household production function and its implications for the allocation of time. *Journal of Political Economy, 83*(2), 255–278.

Rana, S., Parvathi, P., & Waibel, H. (2012, 19–21 September). Factors Affecting the Adoption of Organic Pepper Farming in India. Paper presented at Conference on International Research on Food Security, Germany.

Rogers, E. M. (1958). Categorizing the adopters of agricultural practices. *Rural Sociology, 23,* 345–354.

Ruttan, V. W. (1960). Research on the economics of technological change in American agriculture. *American Journal of Agricultural Economics, 42*(4), 735–754.

Savage, L. J. (1954). *The foundations of statistics.* New York, NY: Wiley.

Schickele, R. (1950). Farmers' adaptions to income uncertainty. *Journal of Farm Economics, 32*(3), 356–374.

Schlaifer, R. (1959). *Probability and statistics for business decisions.* New York, NY: McGraw-Hill.

Schultz, T. W., & Brownlee, O. T. (1942). Two trials to determine expectation models applicable to agriculture. *Quarterly Journal of Economics, 56*(3), 487–496.

Shubik, M. (1952). Information, theories of competition, and the theory of games. *Journal of Political Economy, 60*(2), 145–150.

Shubik, M. (1959). *Strategy and market structure.* New York, NY: Wiley.

Simon, H. A. (Ed.). (1957). A behavioural model of rational choice. In *Models of Man* (pp. 241–260, Chapter 14). New York, NY: Wiley.

Simon, H. A. (1959). Theories of decision making in economics and behavioral science. *American Economic Review, 49*(3), 253–283.

Singh, P. K., Barman, K. K., & Varshney, J. G. (2011). Adoption behaviour of vegetable growers towards improved technologies. *Indian Research Journal of Extension Education, 11*(1), 62–65.

Shackle, G. L. (1949). *Expectation in Economics.* Cambridge University Press.

Shackle, G. L. (1958). *Time in Economics.* North-Holland, Amsterdam.

Shackle, G. L. (1961). *Decision, Order and Time in Human Affairs.* Cambridge University Press.

Strotz, R. H. (1957). The empirical implications of a utility tree. *Econometrica: Journal of the Econometric Society, 25*(2), 269–280.

Swanson, E. R. (1951), Agricultural Resource Productivity and Attitudes to the Use of Credit in Southern Iowa. Iowa State Univ. Dept. Econ. Ph.D. Thesis (Unpublished).

Swanson, E. R. (1957). Problems of applying experimental results to commercial practice. *Journal of Farm Economics, 39*(2), 382–389.

Swanson, E. R. (1959). Selection of crop varieties: An illustration of game theoretic techniques. *Rivista Internazionale di Scienze Economiche e Commerciali, 6,* 3–14.

Thompson, G. L. (1956). Decision making and new mathematics. *Naval Research Logistics Quarterly, 3*(3), 141–145.

Tintner, G. (1942). A contribution to the non-static theory of production. In Lange, O. (ed.), *Studies in Mathematical Economics and Econometrics.* University of Chicago Press.

Throsby, C. D. (1961). Game theory and a time-of-market decision. *Australian Journal of Agricultural Economics, 5*(1), 9–22.

von Neumann, J., & Morgenstern, O. (1947). *Theory of games and economic behavior.* Princeton, NJ: Princeton University Press.

von Neumann, J. and Morgenstern, O. (1953). *Theory of Games and Economic Behavior.* Princeton Univ. Press. (3rd ed.).

Wagner, H. M. (1958). Advances in game theory: A review article. *American Economic Review, 48,* 368–387.

Wald, A. (1950). *Statistical decision functions.* New York, NY: Wiley.

Walker, O. L., & Heady, E. O. (1960). *Application of game theory models to decisions on farm practices and resource use* (Research Bulletin 488). Agricultural and Home Economics Experiment Station, Iowa State University.

Duality Theory of Production

INTRODUCTION

In economic analysis of producer's decision, the objective function and constraints in agriculture are applied jointly. The problem faced by producers in economics is how to optimally allocate their resources, such as land, labour, capital, technology and irrigation. The optimum and efficient use of resources is a challenging task for the producer, especially to maximize his production or minimize cost. Economic theory has recognized that the producer is motivated by the desire to maximize his/her utility or satisfaction. A large number of studies have modelled farm household decision-making behaviour based on the classical theory of firm. All these studies assumed a single objective of profit maximisation as the motivation behind farmers' decision-making behaviour. Thus, these studies have ignored the role of other factors that influence the decisions of farm households, which are usually motivated by multiple, often conflicting goals, rather than profit maximization only. Farming profession remains under pressure due to dependency of climate, and trade-off, which arises between investment in farm activities and family consumption expenditure. In such a situation, the decision-maker usually seeks optimal cooperation among the several objectives (Romero & Rehman, 1989).

In economics, various approaches are used to study the producer behaviour and production analysis, one of which is 'the demand and supply approach', which is based on the parametric demand and supply functions. The parameters are unknown and show the pattern of production which specifies the responses of supply and demand to changes in prices, technology, and scale. Another is the 'production function approach', which is concentrated on the inputs and output relationship. For estimating this relationship, both classical and neoclassical economists have developed different types of production functions and empirically tested them under rigid axioms in different sectors of the economy namely, manufacturing, agriculture and other production systems (Deaton, 1989; Heckelei & Wolff, 2003; Jorgenson, 1984; Just, 1993; Reiss & Wolak, 2007).The most recent approach is the 'duality approach'. The theoretical foundations of duality have been available in economic literature after the seminal work completed by Shephard in 1953, but the empirical applications of duality became more popular later. Extensive empirical research has been made on producer behaviour using the duality approach after the 1960s. In the duality approach, both cost and profit functions are used to calculate elasticity, elasticity of substitution, economies of scale and technical changes. These economic measures provide a deep understanding of the input–output demand and supply models with greater flexibility in the empirical inquiry of the production pattern. This approach has also removed the shortcomings of both demand and supply, and production function approaches.

After the brief review of different economic approaches of the production analysis, it becomes clear that the duality theory has a more flexible approach than the others. The existing theoretical and empirical literatures of the duality theory are presented below along with their findings. Scholars have applied both cost and profit functions for the estimation of economic measures. The use of cost function in production analysis under duality mechanism is made by Chaudhary and Mufti (1999), using cross-section agriculture field survey data at farm household level in Pakistan. There are several other evidences which are used in the United States and other developed countries that can also be applied in the case of India. Therefore, the study has applied cost function approach for estimation of input price elasticity and elasticities of substitutions of both types of farms. The outcomes of empirical estimation of translog cost function under duality framework are presented in the last section.

DUALITY THEORY

The duality approach was developed by R. W. Shephard (1953), while its empirical applications become popular after the 1970s. The first empirical study which exploited the duality theory was made by M. Nerlove (1961) which estimated the Cobb–Douglas cost function as an indirect way of measuring the parameters of the production function of electric utilities. The development of flexible functional forms and its implication to derive plausible functional forms like dual cost and profit functions in the early literature of 1970s. (Christensen, Jorgensen, & Lau, 1973; Diewert, 1971). It was an important step which led to the proliferation of empirical application of the duality theory. There are several studies which are concerned about the agricultural sector. Of these, the study by Binswanger (1974) using US data appears to be one of the earliest.

Why is the application of the duality theory contentiously increasing? The reason behind the growing popularity of the duality theory in the area of production economics is that it allowed greater degree of flexibility in factor demand specification and output supply response equations along with showing a very close relationship between economic theory and practice. For example, suppose transformation or production function depends on a number of input factors, the specified production technology and a vector of output level, then empirical investigation of factor demand equations can be derived first order condition of cost minimization problem. If producer is assumed to be the profit maximizing, then the output supply response equation can also be derived through first order condition of profit function. Unfortunately, very simple restrictive functional forms are used for the function transformation, such as Cobb–Douglas and constant elasticity of substitution (Lopez, 1982). Thus, the use of duality theory permits to side-step problems by solving first order conditions through either directly specifying minimization of appropriate cost function or profit maximization function rather than production.

Theoretical Advances

The theoretical advances of duality pass through various phases, from hypothetical understanding, logical reasoning and mathematical formulation to empirical testing. The empirical application of duality has become popular during 1970s. Probably, the first empirical study of duality theory was made by Nerlove in 1961, which estimated the Cobb–Douglas cost function as an indirect way of measuring the parameters of the production function of electricity utilities. The concept of duality theory is developed as flexible functional form and its applications in the derivation of plausible functional forms, either using cost or profit functions. Diewert (1971) and Christensen et al. (1973) have made early attempts to proliferate empirical application of duality theory. Several of these studies are concerned with the agriculture sector. Of these, Binswanger (1974) study was the earliest in application of duality theory in agriculture by using US data.

The application of duality in production analysis provides an alternative way of solving problems; for example, first order condition by directly specifying suitable maximum profit function or minimum cost function rather than production or transformation functions. In duality mechanism a set of essential properties of profit or cost functions are implied under a 'well-behaved' production technology along with related behavioural assumptions. The duality application has several advantages by specifying profit or cost function rather than transformation function. In order to derive the estimation factors, demand and output supply responses, there is no need to solve any complex production system of the first order condition. The behavioural response equations can be obtained through differentiation of the dual function with respect to input or output prices. Another advantage is its application as it needs less algebraic implications along with the flexibility to specifying complex functions. It does not impose restrictions on the value of elasticities of substitution, separability, homotheticity and so on (Lopez, 1982). During the last four decades, the cost approach has become more popular. It is used to estimate Hicksian input demand in addition to obtaining information regarding properties of the underlying production technology. On the other hand, profit function approach allowed estimating Marshallian factor demands jointly with multi-output supply responses.

THE COST FUNCTION APPROACH

The cost function approach is the most popular approach and is applied for measuring the factor demand elasticities, elasticities of substitution and technical changes in agriculture production. In early literature,

Binswanger (1974) and Kako (1978) specified a translog cost function that estimates factor shares in log linear form. Both had applied the cost function, which was further used by Lopez (1982) as follows:

$$\ln C = \alpha_0 + \alpha_y \ln Y + \sum_i^n v_i \ln P_i + \frac{1}{2}\sum_i\sum_j \gamma_{ij} \ln P_i \ln P_j + \sum_j \gamma_{it} \ln P_i \ln t,$$

(5.1)

where C is the cost of production or cultivation and Y is output, P_i is the price of input factors i and t is used for a time trend variable as a proxy for technical change. Factor share specification can be obtained from equation (5.1) where factors share (S_i) is calculated by using logarithmic differentiation of Shephard's lemma.

$$S_i = v_i + \sum_i \gamma_{ij} \ln P_j + \gamma_{it} \ln t,$$

(5.2)

where $\gamma_{ij} = \gamma_{ji}$ and $i = 1, 2, 3, \ldots, N$

Booth Binswanger (1974) and Kako (1978) measured the elasticities of these inputs: land, labour, machinery, fertilizers and other intermediate inputs. By using the above specification of cost function and share equations, we can separate the effect of biased technical change $(\gamma_{it}$ parameters) of factor shares from the effect of ordinary factor substitution due to change in factor price $(\gamma_{ij}$ parameters) in equation (5.2). The result of both studies shows that technical change is very important and explains ample of the observed changes in factor shares in USA and Japan. Though both the studies were based on the rigid assumption of homothetic production technology with linear expansion paths, changes in the scale of production would not affect factor shares. In the other words, the factor shares in (5.2) are assumed to be independent of output levels. It means that all changes in factor shares are attributed to substitution or factors augmenting technical change. If the production technology is not homothetic, a risk of overestimating the effect of factor substitution or, more likely, technical change exists. It happens because the time trend variable used as a proxy for technical change is generally positively correlated with output levels.

The other study made by Lopez (1980) applied a more general specification of cost function using Canadian agricultural data. This specification allowed for a non-homothetic production function under some degree of flexibility of the translog. In this study he applied a generalized Leontief cost function specification, which is also a flexible functional form of cost function, expressed as follows:

$$C = Y \sum_i \sum_j b_{ij} P_i^{1/2} P_j^{1/2} + Y^2 \sum_i \alpha_i p_i + Y_t \sum_i \gamma_i p_i, \qquad (5.3)$$

In equation (5.3) applying Shepard's lemma, the factor demand equations in input–output ratio can be obtained in following forms:

$$\frac{Xi}{Y} = \sum_j b_{ij} \left(\frac{p_i}{p_j} \right)^{1/2} \alpha_i Y_t + \gamma_{it} t, \qquad (5.4)$$

Where, the coefficient $b_{ij} = b_{ji}$ and $i = 1, 2, 3, \dots, N$

The specification equation (5.4) in Lopez's (1980) analysis allowed separating the effect of relative factor price substitution, factors augmenting technical change and the scale of production on the input–output ratios. Equation (5.4) allows, as a special case, for homotheticity. This occurs if $\alpha_i = 0$ for all i, that is, when the input–output ratios are independent from the output. By estimating a function of four factors (labour, capital, land and structures and other intermediate inputs), input–output ratios showed that the hypothesis of homotheticity is rejected by a wide margin and that changes in the scale of production explain a very important proportion of changes in the input–output or share equations. The effect of non-neutral technical change was found to be insignificant, which was a rather surprising result. However, a recent more disaggregated study by Lopez and Tung (1982) using combined cross-section and time series data for Canadian agriculture and considering inputs: energy, energy-based, labour, capital, land and other intermediate inputs, shows that the factors augmenting technical change parameters (*it*) were jointly significant. Though the technical change effect was substantially less dramatic than those obtained by Binswanger (1974) and Kako (1978), the output scale effect is very strong and significant.

Table 5.1 shows that the own factor price elasticities of Hicksian input factors demand are quite similar for the four studies, despite using different data and models. The results have concluded that

Table 5.1 *Hicksian Input Demand Elasticities*

Study	Data	Prod. Function	Finding (Input DE)
Binswanger (1974)	US Ag; cross sec + time series	Translog	−0.34 (L), −0.91 (La), −0.95 (Fer + Ch), −1.09 (K)
Kako (1978)	Japan rice farm, cross-section + time series (1953–1970)	Translog	−0.49 (L), −0.46 (La), −0.32 (Fer + Ch), −0.59 (K)
Lopez (1980)	Canada Ag; time series (1946–1977)	Gen Leontief	−0.42 (L), −0.52 (La), −0.41 (Fer + Ch), −0.35(K)
Lopez and Tung (1982)	Canada Ag; cross sec + time series (1961–1979)	Gen Leontief	−0.42 (L), −0.39 (La), −0.89 (Fer + Ch), −0.63(K)

Source: Lopez (1982).
Note: L = land; La = labour; Fer+Ch = fertiliser and chemical and K = farm capital.

factor demands are inelastic; where land (L) demand elasticity ranges from −0.35 to −0.50, the demand of labour (La) elasticity ranges from −0.40 to −0.50, but the Binswanger's result presents an outlier. The demand for fertilizers and chemicals (Fer + Ch) tends to be more elastic at least in the studies using North American data (−0.9) and farm capital (K) demand also exhibits somewhat lower values than the former. It means the estimated demand elasticities may provide some guidance to policy makers with several notions of the various degrees of price responsiveness of the inputs used in agricultural production.

These studies have applied different cost functions and found that the input demands are moderately responsive to prices. There exists a significant substitution possibility among several input pairs of which energy-based inputs and land appears to exhibit the greatest potential. The aggregate agricultural technology is not homothetic and the simpler production function specifications such as the Cobb–Douglas or Leontief are not appropriate specifications as shown by the studies by Binswanger (1974a) and Lopez (1980), respectively.

THE PROFIT FUNCTION APPROACH

All the empirical literature of duality theory is based on the cost function approach, having a common limitation that the output technology is homothetic or non-homothetic. Another, serious limitation of the cost approach is that it assumes that output levels are not affected by changes in factor price, hence the indirect effect of change in factor price on factor demand remains unnoticed. Along with the above, the insertion of output levels as explanatory variables may lead to simultaneous equation biases if output levels are not exogenously determined. The above discussed problems certainly become more complex if a multi-output cost function is estimated (McKay et al., 1983; Thompson & Langworthy, 1989). In that condition the input shares or input–output ratios are normally used to estimate the factor demands and are dependent on each of the outputs when constant returns to scale are assumed (Hall, 1973). These share equations are non-linear in the various outputs, making it very difficult to use econometric techniques designed to tackle simultaneity problems (Lopez, 1982).

The profit function approach allows overcoming of those limitations which are assumed in using cost function and conquer to strengthen behavioural assumption. The profit function approach and its maximization assumption may be substantially more difficult to support in agriculture than simple cost minimization because of risk-related problems which are largely related to the variability of output yields and price rather than to the cost of production. The factors demand can be estimated by using a profit function framework which allows measuring input substitution and output scale effects of change in factor price. In addition, the cross effects of output price changes on factor demands and vice versa as well as output supply responses and their cross-price effects are also measured. A major gain of using profit function approach is that it allows for the estimation of multi-output technologies in a much simpler way than a cost function or a transformation function. The profit function can be written as follows;

$$\Pi(p, w; K) = \{\text{Max}_{yw}\, ywp\, y - w\, x : F(y, x; K) = 0\}, \qquad (5.5)$$

where y is a vector of M outputs, x is a vector of N variable inputs, K is a vector of S fixed inputs $F\,(\cdot)$ is a continuous, concave transformation function, p, w, are vectors of M output prices and N input prices. The profit function $\Pi\,(\cdot)$ is non-decreasing in p, non-increasing in w, linear homogeneous and convex in p and w. Hessian matrix is symmetric with respect to p and w. As in the case of the cost function, its properties allow developing suitable functional specifications which permit to test, verify or impose the above properties. The factors demand and output supply equations are derived from the specified profit function by simple differentiation with respect to input prices and output prices, respectively (Hotelling's lemma). Furthermore, the shadow price of fixed resource K_i is the derivative of $\Pi\,(\cdot)$ with respect to K_i.

The majority of literature on profit function has assumed a single output technology. The earlier work of Lau and Yotopoulos (1972) and Yotopoulos, Lau and Lin (1976) have used a Cobb–Douglas specification for a single output restricted profit function. They estimated output supply and input demand responses using data from India and Taiwan, respectively. Other studies have applied flexible profit function approach. Binswanger and Evenson (1984) tried various single output flexible form specifications using Indian data including the generalized Leontief, translog and the quadratic normalized function. They found that using the translog specification was less compatible with the restrictions implied by economic theory than the other two forms. An undesirable feature of the specification used by Binswanger and Evenson for both the generalized Leontief and normalized quadratic forms of profit function is that the shadow prices of fixed resources implicitly assume constant, independent of the level of fixed resources.

Sidhu and Baanante (1981) applied profit function approach to analyse the demand of inputs and supply of wheat in Punjab state in India. The normalized restricted translog profit function was used for considering wheat output, three variable inputs (labour, fertilizers and animal power) and seven fixed factors (machinery, equipment, land, various soil nutrients, schooling and irrigation area). They obtained estimates for the elasticities of wheat supply responses as well as for the three-variable factor demands. They also used the Cobb–Douglas type profit function, but it is not supported by the data, and that the symmetry restrictions are not rejected. They obtained a wheat supply elasticity of 0.6 and, surprisingly, they found that the output price effect is more powerful in affecting demand for labour, fertilizers and animal power than their respective prices.

Lopez (1981a) applied a multi-output profit function to estimate agriculture input and output parameters. The estimation reports on two outputs (animal and crop), four inputs (land, capital, hired labour and operator labour) by using generalized Leontief profit function for cross-sectional Canadian census data. The findings suggest that hired labour and operators complement each other rather than being substitute inputs. His study has some specific features: (a) A simple test for economies and diseconomies of joint production; (b) The hypothesis of no rejection of non-joint production of crop and animal outputs. It is accepted at a given high level of output aggregation, but at a more disaggregated level the hypothesis may be rejected; (c) A procedure to separate expansion and substitution effects for both inputs and outputs from the profit function estimates. This method helps to derive the output trade-offs due to a change in one output price for given level of inputs, and to measure the input substitution stemming from a factor price change for a given output level; (d) The Marshallian elasticities are obtained directly by the parameter estimates of the profit function and the trade-offs along the production possibility frontier and isoquants; (e) The consideration of hired and operator or family labour as two different inputs. The Marshallian estimates indicate that hired and family labour respond very differently to changes in output and input prices, and thus are regarded as different inputs. Generally, hired labour is more responsive to price changes than family labour.

The application of the profit function approach has allowed researchers to develop relatively simple tests for the existence of allocative and technical efficiency of farm production primarily in developing countries. Lau and Yotopoulos (1971) applied a Cobb–Douglas profit specification to test for relative efficiency of Indian producers; a number of studies have used similar approaches. For example, Sidhu and Baanante (1979) using Punjab state data found that the producers do obtain allocative efficiency and that the profit function approach appears to be an appropriate concept in the analysis of factor demand and output supply responses.

FARM HOUSEHOLD APPROACH (MICRO APPROACH)

The application of duality theory at household level data uses both linear and non-linear objective functions. The linear function is found in case of time series and pool time series data (Binswanger, 1974; Koko 1978; Lopez, 1980; Lopez & Tung, 1982). While non-linear objective function is applied by Epstein (1981), Lopez (1981), Barnum & Squire (1980), Hymer & Resnick (1969) and Sen (1966). Lopez (1881) demonstrated that the farm household optimization problem can be maximized using non-linear utility function subject to a non-linear budget constraint. The constraint is non-linear for the reason that a significant proportion of farm household income is obtained from the farm returns, which is also a non-linear function of household labour and fixed factors of production. The generalized model of farm household is applied by Sen (1966) for the production analysis. He assumed no off-farm employment, and that all outputs and inputs have exogenous prices and only labour (shadow) prices are determined exogenously by farm household complex. Further, an assumption was made by Lopez (1982) that the farm household utility function is well-defined as a function of leisure and the consumption vector of market purchased goods and services. The utility maximization problem of farm household is written as;

$$\text{Max}_{T-W, X} \, U \, (T - W, X) \qquad (5.6)$$

$$a. \; p \, X \leq \Pi \, (p'; L, W) + y$$

$$b) \; T \, X \geq T - W \geq 0, X \geq 0$$

Where $U \, (\cdot)$ is the farm household utility function, T is total number of hours available for household members for work and leisure, p represents a vector of N market purchased goods and services. X is a vector of N market purchased consumption goods and services. W is the number of working hours, $\Pi \, (\cdot)$ is a conditional variable profit function, p' is an exogenous price vector of S net outputs produced by farm, using the rule of representing output prices by positive sign and purchased input pieces negative sign, L is the fixed factor of production, say land and y is the non-labour income net of fixed requirements.

Equation (5.6a) shows that the total expenditure on consumption goods cannot exceed the income associated with net farm returns to labour and owned fixed resources represented by profit plus net non-labour incomes that include government transfers, assets income and so on. The conditional variable profit function Π $(p'; L, W)$ of farm household is defined as;

$$\Pi\ (p';\ L,\ W) \equiv \text{Max}\ Q\ \{p';\ Q\colon (Q;\ L,\ W)\ \varepsilon\tau\} \tag{5.7}$$

Where Q is the vector of S set net output including M outputs and $S-M$ inputs, τ is the production possibilities set based on the assumption of compact, non-empty, convex set. Therefore, the verification of Π $(p';\ L,\ W)$ is non-negative, continuous linearly homogeneous and convex in p', non-decreasing and concave in L and W for fixed p'.

If the problem (5.6) is defined locally for the compact subset M, then it can be also defined via an indirect utility function associated with a problem is following method;

$$G\ (p,\ p';\ L,\ y) \equiv \text{Max}_{T-W,\ X}\ \{U(T-W,\ X)\} \tag{5.8}$$
$$a)\ p\ X - \Pi\ (p';\ L,\ W) \leq y$$
$$b)\ (T-W,\ X)\ \varepsilon f\ \text{and}\ \{(p,\ p';\ L,\ y)\ \varepsilon p\}$$

Where the attention is constrained, the set of utility levels $M \equiv \{\mu\colon \mu' \leq \mu \leq \mu''\}$ implies that the correspondent commodity space f and parameter space p are compact, non-empty set.

Epstein (1981) has illustrated the duality relationship in the context of a more general non-linear model in equation (5.8) and applied indirect utility function $G\ (\cdot)$ that shows one to one duality relationship between the $G\ (\cdot)$ and $U\ (\cdot)$ for a given profit function $\Pi\ (\cdot)$. Here, $G\ (\cdot)$ is non-increasing function of p, non-decreasing of p', L and y and homogeneous of degree zero in p, p' and y. In addition, minimisation of $G\ (.)$ subject to the budget constraint allows retrieving a function U^* with identical behavioural implications as U and, from the first order conditions of this minimisation problem, a relationship can be obtained through partial derivation between the indirect utility function and the farm-household behavioural equations:

$$Q_i = \frac{\partial G / \partial P_i'}{\partial G / \partial y} I = 1,\ldots,S \qquad (5.9a)$$

$$X_i = \frac{\partial G / \partial P_j}{\partial G \partial y} j = 1,\ldots,N \qquad (5.9b)$$

$$\frac{\partial \Pi}{\partial T} = \frac{\partial G / \partial T}{\partial G / \partial y} \qquad (5.9c)$$

From the equations (5.9a) it is clear that the output supply is dependent on the structural properties of both the production technology and household's preferences. In addition, output supply responses are affected by the price level of those commodities consumed by the household as well as by the household's non-labour income, and not by the level of net output prices. In the same way, demand for consumer goods is jointly determined by the parameters of the consumption and production sides. The equations specified by (5.9a) are unconditional; in the sense that they are evaluated at the utility maximising level of W. Equation (5.9c) provides a specification for the shadow price of land $\partial \Pi / \partial T$. If the conditions of the implicit function theorem are satisfied by $\partial \Pi / \partial T$ (p', W, L) then it can be derived from (5.9c) a specification for the equilibrium utility maximising level of work, W.

Further, Lopez (1981b) presented the farm household consumption demand functions; according to him, the net output supply functions along with the equilibrium level of hours of work are homogeneous of degree zero in consumption good prices, net output prices and non-labour incomes. Generally, the net output supply functions of the conventional firm, the net output supply functions of the farm-household are not homogeneous of degree zero in net output prices. Moreover, contrasting the conventional model, the farm-household consumption decisions (i.e., demand for consumer goods), production decisions (i.e., demand for production factors) and the equilibrium level of work are all dependent on the parameters of both the demand

and supply side model. It means the conventional model of firm; net output supply functions are not necessarily upward sloping. Although, Lopez has derived a compensated net output supply expression, which shows to be non-negative in the following farm:

$$\frac{\partial Q_i}{\partial p_i'} - \frac{\partial Q_i}{\partial W} \cdot \frac{\partial W}{\partial y} Q_i = e_{p_i'p_i'} \geq 0 \qquad (5.10)$$

$$i = 1, 2, 3 \dots S$$

Where $e_{p_i'}$ is the second partial derivative with respect to p', of an expenditure function $e(p, p', T; \mu)$ defined by

$$e(p, p', T; \mu) = \mathrm{Min}_{T-w, x} \{px - \Pi(p'; T, W) : U(T - W, x) \geq \mu\} \qquad (5.11)$$

Hence, the sign of the directly observed Marshallian output supply effect $\partial Q_i / \partial p_i^j$ cannot be predicted, the compensated or 'Hicksian' output supply effect (i.e., the left-hand side of equation (5.11) is non-negative. Thus, the estimate $\partial Q_i / \partial p_i' - \partial Q_i / \partial W$ and $\partial W/\partial y$ it is possible to empirically verify inequality via equation (5.11). This additional testable prediction is obtained from the farm-household model. One more prediction from the model is that, although the effect of a change in net output price p_i' on the equilibrium level of work W is in general unknown, the utility constant effect is unambiguously non-negative.

$$\partial Q_i / \partial W \{\partial W/\partial p_i' - \partial W/\partial y Q_i\} = -(e_{p_i'p_i'} + \Pi_{p_{ii}'}) \geq 0 \qquad (5.12)$$

$$i = 1, 2, 3, \dots, S$$

Hence, if the equation (5.9) is estimated subsequently the left-hand-side of (5.12) can be calculated and the non-negative restriction in equation (5.12) that can be tested. It is important to note that the sign of $\partial W/\partial p_i'$ is in general vague and that if the weak assumption that $\partial Q_i/\partial W > 0$ is made, then a testable prediction of the model is

$$\frac{\partial W}{\partial p_i'} - \frac{\partial W}{\partial y} Q_i \geq 0, \text{for } i = 1, 2, 3, \dots, S$$

Symmetry of the function e allows, for showing that reciprocity conditions between production and consumption decisions also hold. Lopez (1981, 1982) showed the subsequent testable symmetry relationship, which is obviously absent, in the conventional models of the firm and the household:

$$\partial x_i^* / \partial p_j^i = -\partial Q_j^* j / \partial y (p'; T, W^*) / \partial p_i \qquad (5.13)$$

For all $i = 1, 2, 3, \ldots, N$ and $j = 1, 2, 3, \ldots, S$.

Where the compensated demand effect $\partial x_i^* / \partial p_j' = \partial x_i / \partial p_j' - \partial x_i / \partial y Q_j$ and the compensated outputeffect of a change in consumption goods price $\partial Q_j^* / \partial p_j = \partial Q_j / \partial W \{ \partial W / \partial p_i + \partial W / \partial y\, x_i \}$. Thus, symmetry relations presented in the equation (5.19), in terms of expressions which can be empirically estimated and testable prediction of the farm-household model.

Similar results to (5.10), (5.12) and (5.13) can be derived for the case in which farmer's works off-farm and when the farm-household produces goods, which are entirely consumed within the farm-household (Lopez, 1981, 1982). If household members work on off-farm practices then important questions arise that whether they regard on-farm and off-farm work as perfect substitutes in consumption and if there exist binding restrictions on the number of hours, which they can work off-farm. If they regard on-farm and off-farm work as identical 'commodities' and if they face no binding restrictions on hours of off-farm work, then the shadow price of labour becomes exogenous. If, in addition, all outputs produced by the farm-household are at least partially traded, then possibility arises to dichotomise production and consumption decisions. In such a condition, the conventional models of the firm and farm household can be applied, and the prediction discussed above is no longer possible to apply. However, if any of the above conditions are not met, then utility maximising and profit maximisation decisions are interdependent and follows Lopez previous analysis.

CONCLUSION

The early work of duality theory is conducted by Shephard in 1953, after that a series empirical studies are conducted in 1960s. The use of duality theory developed as advance method of estimation factors elasticity of substitution and price elasticity of demand use cost function and profit function instead of production function. The result of duality method estimation is more precise than the method of input output production function. In both method of Input-Output and duality approach of Cobb-Douglas, CES and translog production function are used as techniques of estimation, but the result is more precise in case duality theory.

REFERENCES

Barnum, H. N., & Squire, L. (1979). *A model of an agricultural household: Theory and evidence.* World Bank staff occasional papers; no. 27, USA.

Binswanger, H., & Evenson, R. (1984). Labor demand in North Indian agriculture. In Binswanger and Rosenzweig (eds.), *Rural labor markets in Asia: Contractual arrangements, employment and wages.* New Haven: Yale University Press.

Binswanger, H. P. (1974). A cost function approach to the measurement of elasticities of factor demand and elasticities of substitution. *American Journal of Agricultural Economics, 56*(2), 377–386.

Chaudhary, M. A., & Mufti, S. S. (1999). Translog cost function estimation of farmer production and employment relationships. *Pakistan Economic and Social Review, 37*(1), 39–60.

Christensen, L. R., Jorgenson, D. W., & Lau, L. J. (1973). Transcendental logarithmic production frontiers. *Review of Economics and Statistics, 55*(1), 28–45.

Deaton, A. (1989). Household survey data and pricing policies in developing countries. *The World Bank Economic Review, 3*(2), 183–210.

Diewert, W. E. (1971). An application of the Shephard duality theorem: a generalized Leontief production function. *The Journal of Political Economy, 79*(3): 481–507.

Epstein, L. (1981). Generalized duality and integrability. *Econometrica, 49*(3), 655–678.

Hall, R. E. (1973). The specification of technology with several kinds of output. *The Journal of Political Economy, 81*(4), 878–892.

Heckelei, T., & Wolff, H. (2003). Estimation of constrained optimisation models for agricultural supply analysis based on generalised maximum entropy. *European Review of Agricultural Economics, 30*(1), 27–50.

Hymer, S., & Resnick, S. (1969). A model of an agrarian economy with non-agricultural activities. *The American Economic Review, 59*(4), 493–506.

Jorgenson, D. W. (1984). *Econometric methods for modeling producer behavior.* Cambridge, MA: Harvard Institute for Economic Research.

Just, R. E. (1993). Discovering Production and Supply Relationships: Present Status and Future Opportunities. *Review of Marketing and Agricultural Economics, 61*(1), 1–30.

Kako, T. (1978). Decomposition analysis of derived demand for factor inputs: The case of rice production in Japan. *American Journal of Agricultural Economics*, 60(4), 628–635.

Lau, L. J., & Yotopoulos, P. A. (1971). A test for relative efficiency and application to Indian agriculture. *The American Economic Review*, 61(1). 94–109.

Lau, L. J., & Yotopoulos, P. A. (1972). Profit, supply, and factor demand functions. *American Journal of Agricultural Economics*, 54(1), 11–18.

Lopez, R. E. (1980). The structure of production and the derived demand for inputs in Canadian agriculture. *American Journal of Agricultural Economics*, 62(1), 38–45.

Lopez, R. E. (1981). *Testable implications of the farm-household model* (Paper No. 81–02). Department of Agricultural Economics, University of British Columbia.

Lopez, R. E. (1982). Applications of duality theory to agriculture. *Western Journal of Agricultural Economics*, 7(2), 353–365.

Lopez, R. E., & Tung, F. L. (1982). Energy and non-energy input substitution possibilities and output scale effects in Canadian agriculture. *Canadian Journal of Agricultural Economics*, 30(2), 115–132.

McKay, L., Lawrence, D., & Vlastuin, C. (1983). Profit, output supply, and input demand functions for multiproduct firms: The case of Australian agriculture. *International Economic Review*, 24 (2), 323–339.

Nerlove, M. (1961). *Returns to scale in electricity supply.* Stanford, CA: Institute for Mathematical Studies in the Social Sciences.

Reiss, P. C., & Wolak, F. A. (2007). Structural econometric modeling: Rationales and examples from industrial organization. J. Heckman and E. Leamer (eds.). In *Handbook of econometrics* (Vol. 6, pp. 4277–4415). Elsevier, North Holland.

Romero, C., & Rehman, T. (1989). *Multiple criteria analysis for agricultural decisions.* Developments in Agricultural Economics (Netherlands), volume no 11. Elsevier, Amsterdam, 1989 and second edition in 2003.

Sen, A. K. (1966). Peasants and dualism with or without surplus labor. *The Journal of Political Economy*, 74(5), 425–450.

Shephard, R.W. (1953). *Cost and production functions.* Princeton, NJ: Princeton University Press.

Sidhu, S. S., & Baanante, C. A. (1979). Farm-level fertilizer demand for Mexican wheat varieties in the Indian Punjab. *American Journal of Agricultural Economics*, 61(3), 455–462.

Sidhu, S. S., & Baanante, C. A. (1981). Estimating farm-level input demand and wheat supply in the Indian Punjab using a translog profit function. *American Journal of Agricultural Economics*, 63(2), 237–246.

Thompson, G. D., & Langworthy, M. (1989). Profit function approximations and duality applications to agriculture. *American Journal of Agricultural Economics*, 71(3), 791–798.

Yotopoulos, P. A., Lau, L. J., & Lin, W. L. (1976). Microeconomic output supply and factor demand functions in the agriculture of the province of Taiwan. *American Journal of Agricultural Economics*, 58(2), 333–340.

Market Structure

Two Group Non-collusive Model

INTRODUCTION

In economics the concept of collusive model falls under the oligopoly market structure where the number of competitors is few and products are differentiated. The concept of group was introduced by Chamberlin in 1933 for the analysis of imperfect market firm's output and price decision behaviour. His concepts of both collusive and group are used for the theoretical analysis of organic inputs and output market. Why is there a need for theoretical explanation of the organic market? The answer to this question is to grow the demand of organic inputs and output market across counties. Why is there a need for organic farming? This is because unconscious use of chemical fertilizers and pesticides have damaged the fertility of land and water resources. Along with the above, several ill effects of conventional farming have been noticed on environment and human health.

The demand for organic products is created by the middle and upper classes, while the international demand of organic products is related to the quality of products. So the aggregate demand of organic products inspired producers to grow crops organically. It has been found in empirical research that both domestic and international organic consumers are willing to pay higher prices for organic products due to their rich nutritional ingredients. The export of organic products is foreign-demand driven. So the demand and supply of organic products will determine the price of organic products, but the market structure may be oligopolistic in nature where the producers are few but buyers are large in number.

The different types of market structure are used to explain producer and consumer behaviour in economic theory. But here oligopoly market structure is chosen for theoretical interpretation of organic and non-organic input output market. Empirically it has been found that the market plays a crucial role in the growth and development of any new industry. Thus, the necessary condition for the growth of the organic industry is market facility that will be facilitated by the state. Quality check is essential in case of organic products so that they meet international standards (Anon, 2005a). India set up its own quality standards based on international norms and parameters (NPOP, 2005). Quality standard provisions are not similar for domestic and international markets. The domestic consumers are consuming organic products without observing the quality standards. On the supply side, the growth and development of the organic inputs market is equally important. Here we assume that the organic inputs and output market has imperfect competition or oligopoly market structure and within this market structure consider two group producer model in organic output market. The upstream input supplier sells their inputs to both type farmers.

Thus, the growth of organic output market depends on the certified area and productivity per hectare, where the certified area is directly related with the decisions of farmers, while the productivity of land depends on the various factors, and technology is one of them. In India majority of farmers are using traditional farming methods, such as using domestic seeds and compost for cultivation is very close to organic or natural farming. Although the Green Revolution have made tremendous improvement in seeds, fertilizer and methods of cultivation. The use of chemical fertilizers and pesticides along with high-yielding varieties of seeds lead to increase in production and productivity, but at the same time polluted water and degraded soil fertility (Bhattacharyya, 2015; Rao et al., 2014; Saleh et al., 2014).

Degradation of natural resources and human health due to damage of ecosystem pushes people to back to the nature. Civil society, international organizations, NGOs, environmentalists and governments are engaged in spreading awareness among the people. Agriculture production activity is directly degrading water, soil and ecosystem using unsustainable methods of cultivation. So the alternative method of organic farming is getting space among the farming community, but is limited to educated and large farmers only. This limitation is due to high accreditation and certification cost, low productivity, access to market facility and financial constraints. To eliminate these limitations, a clear-cut organic farming policy is made by the state.

This chapter has three subsections. The first section describes an economic model designed for organic and non-organic producers under the imperfect market condition. The second section explains product differentiation and shape of demand and cost curves for both types of products. The third section presents the theoretical formulation of the two group-two price model based on theory of firm under oligopoly non-collusive market structure is explained. Concluding remarks are presented in the end of the chapter.

TWO GROUPS NON-COLLUSIVE MODEL

The problem in organic production analysis is choosing what crop choose to produce, which method to use for production and for whom to produce? All these are central issues in production economics which are related to the decision behaviour of a producer. In a market-led economy these decisions are determined by prices of organic inputs and output. However, prices of organic inputs and output are determined through interaction of demand and supply schedules in respective markets. Both inputs and output market of an organic market are linked with the money market, which facilitates the capital for production.

In economics the product market is classified on grounds of competitiveness and product classification. There are two extreme market structures found theoretically, that is, monopoly and perfect competition. But in real life both are rarely found. Between these two extreme markets, imperfect competition arises and within this oligopoly exists. If there is a large number of firms that produce homogeneous products, then it is called perfect competition; if there is a single producer, then it is called monopoly; and if there are few producers that produce differentiated products, then it is called imperfect competition. In perfect competition, the price is determined by the industry and firm can follow it; in monopoly firm is itself the industry and decides price itself; and in imperfect market, especially in oligopoly, price is decide mutually (Knight, 1933).

Assumptions

1. The goals of producer are multiple and profit maximization is one of them.
2. The whole economy is divided into two groups: organic and non-organic.
3. Each group has a large number of consumers and producers; the larger group has a large number of consumers and producers than the smaller group.
4. Free entry and exit is allowed to the consumers and producers within the group as well as across the groups.
5. Both cost and demand curves of producers are uniform within each group.
6. There are two sets of equilibrium output and price: one for the organic group and another for the non-organic group
7. Prices of factors and technology are given for both groups, but they are different from each other.
8. The long run consists of a number of short-run periods; each time period is independent in terms of technology choice.

Cost Curves

The classical economists tried to solve the issues of cost and demand curves' shapes using their wisdom. Firstly, they introduced a downward-falling demand curve for individual firms. Secondly, they adopted a general equilibrium approach for shifting of cost curve using the concept of external economies of scale. Finally, they introduced a U-shaped selling cost curve in the model. All these classical provisions were adopted by Sraffa (1926) who presented a model for individual firms where the demand curve had a negative slope. His result of demand curve slope was both theoretically and empirically sound and got universal acceptance.

Furthermore, Joan Robinson (1933) and E. Chamberlin (1933) adopted his views in their works individually and found the same result for a firm despite using different methodological and analytical tools (Koutsoyiannis, 1994). They introduced the concept of group in producer analysis. The products produced within the group are homogeneous, but they are different from those of other groups.

The concept of two groups has been adopted from their works. The differentiation in products may be due to quality, packing, brand, features and production technology. These products are sold in the same market at different prices. For example, mobile manufacturing companies produce differentiated products, that is, mobiles and sell them in the same market at different prices. The same method is adopted for analysis of organic products.

The cost curve used in the model follows Chamberlin's type of firm. Average total cost (ATC), MC and AVC have the usual U-shaped curves. It is assumed that all individuals produce only a single crop in one cultivation season and produce optimally without any natural risk. It is further assumed that the cost of advertising, packaging and processing is borne by the selling agency, while certification and cultivation cost is borne by farmers. The government may authorize a third party agency for product quality accreditation and certification of land.

PRODUCT DIFFERENTIATION, DEMAND AND COST CURVE

While reviewing economic literature on falling demand and cost curves, we found that Sraffa had introduced the concept of product differentiation in the analysis of firm behaviour. However, initially Chamberlin had applied the concept of product differentiation in price and output determination of firm, along with advertainment cost. He pointed out that the demand of a product is not only determined by price but also by associated services and selling strategy of the firm. In case of some products the demand depends on the method of production. For instance, the products which are produced using child labour are banned in developed countries, similarly organic products are not sold without accreditation. In the analysis of firm, Chamberlin added two policy variables: one was selling activities and the other was the product itself.

The cultivation method marks the difference between organic and conventional products. The price and output determination are similar for organic and conventional products under the non-collusive two group oligopoly type market structure. Here selling activities are not associated with the producer and are not included in the model, but policy variable is included as the selling strategy belongs to the selling agents. Thus, the demand of products is associated with cultivation method, certification and price. The shifting of demand curves depends on the following:

1. Change in the quantity, method of cultivation, quality and accreditation
2. Number of producers in each group, price of product and price of its substitute
3. Teste preference, income of consumers and selling strategy of agents (Koutsoyiannis, 1994).

Here the differentiation of products is due to differences in method of cultivation of crops. Each producer within a group produces homogeneous products, that is, crops using the same method of production. So this kind of product differentiation has inherent characteristics which are due to use of different inputs and farming methods in cultivation. For example, wheat is produced by both groups of producers, but its inherent nutritional content and taste differs due to differences in method of cultivation and use of different kinds of seeds and fertilizers.

The quality assurance of organic products that is done by accreditation provides confidence among the consumers and attracts them to buy the products. Fancied differentiation can be made through packaging, brand name and advertising. Koutsoyiannis (1994) pointed in his book that differentiated products are close substitutes of each other, having high value of cross-elasticity of demand. The difference in price of organic and conventional products is due to the inherent product differentiation. Consumers are ready to pay a higher price for organic products due to better quality, better taste and heath concern. Both organic and conventional groups sell their products at different prices and face kinked demand for each group of consumers. The productivity under organic cultivation is lower than conventional one, so if the organic group is not able sell at a higher price, then this group will face a loss and as a result, a producer of the organic group exits and joins the conventional group.

TWO GROUP-TWO PRICE MODEL

The idea of two group-two price model comes from the work of Chamberlin and Paul Sweezy in 1993. Here two demand curves are presented: one for conventional products and another for organic products. At the same time two prices are determined in the same market: One for conventional products and another for organic products. The two prices exist in the market because of demand and supply constraints of both types of products.

The consumers are divided into two groups: one small group for organic products and other large group for conventional products. On the producer or supply side, the price of organic products is higher than conventional products in the market due to supply, productivity and cost constraints. On the consumer and demand side, organic consumers are ready to pay a higher price for organic products due to environmental concern and better nutritional quality. As found in empirical studies, organic producers grow the same crop at a lower price than the conventional producers, but the productivity is lower. Hence, they sell the crop at a higher price to organic consumers.

In this model, it is assumed that the agriculture industry has two kind of producers and consumers. This itself leads to forming of two groups for producers and two for consumers. The organic group of producers produce the same crops using organic farming methods while conventional producers use conventional farming methods. Both types of producers sell their own products in the market without collusion. Here the groups as a whole have the decision-making power related to determination of price rather than an individual producer within the group. The group of organic producers is smaller than conventional producers and this is similar for the groups of consumers. The small group of organic producers supply a very small part of the total market demand, whereas the larger group of conventional producers supply the remaining large section of the total market demand at a lower price.

Figure 6.1(a) shows that OX_0 is the amount of organic product that organic producers are selling at a higher price (OP_o) than the conventional product price of and receiving $aebP_o$ amount of profit.

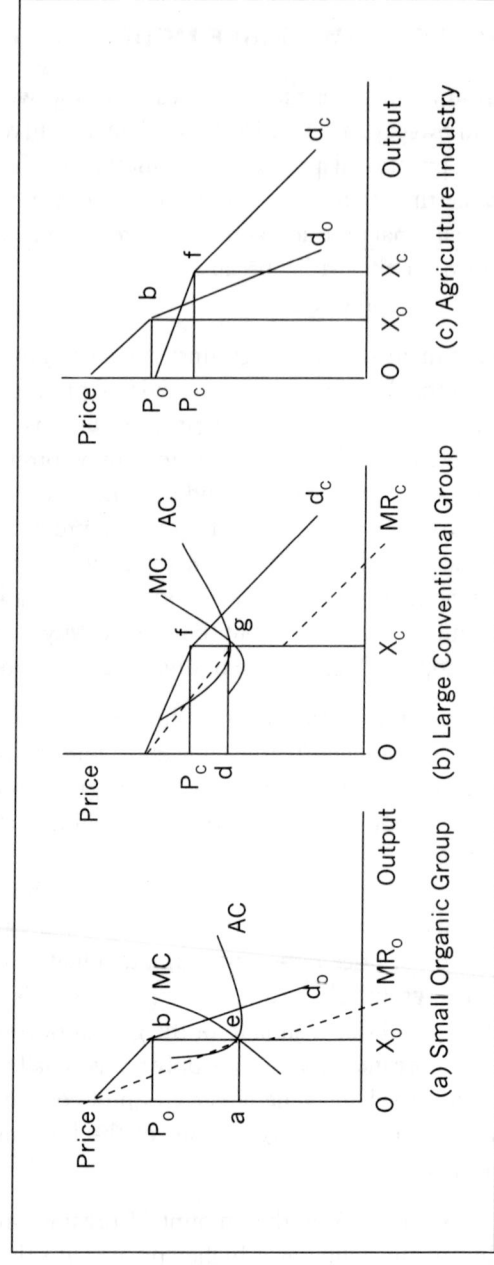

Figure 6.1 *Price and Output Determination in Product Market*

Similarly, Figure 6.1(b) shows that conventional producers are supplying a large amount of conventional product OX_c at a lower price OP_c to larger group of conventional consumers and earning $dgfP_c$ amount of profit. Figure 6.1(c) presents the agriculture sector market demand and prices for both types of products using kindled demand approach.

CONCLUSION

This chapter tried to develop a theoretical model of two-group, two-commodities and two-market model for organic and non-organic consumer and producers. This model is based on certain assumptions and followed the approach of Chamberlin and Paul Sweezy's (1993) two-group model. But the two group models of Paul Sweezy and Chamberlin were criticized on the same ground of price rigidity. In the same line of price rigidity criticism, we proposed two-price levels and two demand curves: one for organic products and the other for convention products in the economy. The consumers of organic consumers are willing to pay a higher price for organic produce due to better quality of products and environmental concerns.

REFERENCES

Aanon. (2005b). Import plant quarantine requirements of fresh sweet orange fruits from the Republic of South Africa. Online at: http:/www.nda.agric.za/docs/npposa/Koreacitrusprotocol.htm

Bhattacharyya, R., Ghosh, B. N., Mishra, P. K., Mandal, B., Rao, C. S., Sarkar, D., … & Franzluebbers, A. J. (2015). Soil degradation in India: Challenges and potential solutions. *Sustainability*, 7(4), 3528–3570.

Chamberlin, E. (1933). *Theory of Monopolistic Competition*. Cambridge, MA: Harvard University Press.

Cohen, A. J. (1983). 'The Laws of Returns under Competitive Conditions': Progress in Microeconomics since Sraffa (1926)? *Eastern Economic Journal*, 9(3), 213–220.

Giller, K. E., Beare, M. H., Lavelle, P., Izac, A. M., & Swift, M. J. (1997). Agricultural intensification, soil biodiversity and agro ecosystem function. *Applied Soil Ecology*, 6(1), 3–16.

Jaeck, M., Lifran, R., & Stahn, H. (2012). *Emergence of Organic Farming under Imperfect Competition* (No. 1239). Aix-Marseille School of Economics, Marseille, France.

Koutsoyiannis, A. (1994). *Modern Microeconomics*. Macmillan

Knight, F. H. (1933). 'Risk, Uncertainty and Profit'. London School Reprints of Scarce works, No. 16.

Rao, M. A., Rizvi, S. S., Datta, A. K., & Ahmed, J. (Eds.). (2014). *Engineering properties of foods*. CRC press.

Robinson, J. (1933). *The Economics of Imperfect Competition*. Macmillan

Saleh, R., Robinson, E. S., Tkacik, D. S., Ahern, A. T., Liu, S., Aiken, A. C., ... & Donahue, N. M. (2014). Brownness of organics in aerosols from biomass burning linked to their black carbon content. *Nature Geoscience*, 7(9), 647.

Singh, R. B. (2000). Environmental consequences of agricultural development: A case study from the Green Revolution state of Haryana, India. *Agriculture, Ecosystems & Environment*, 82(1), 97–103.

Sraffa, P. (1926). The law of returns under competitive conditions. *Economic Journal*. 36(144), 535–550.

Sweezy, P. M. (1939). Demand under conditions of oligopoly. *The Journal of Political Economy*, 47(4), 568–573.

Wandel, M., & Bugge, A. (1997). Environmental concern in consumer evaluation of food quality. *Food Quality and Preference*, 8(1), 19–26.

Producer Behaviour and Niche Market

INTRODUCTION

Producer behaviour in adopting organic farming is influenced by social, economic, climatic, market and policy factors. But the market is the only factor that determines the prices, whose outcome is profit for the producer. The term 'niche market' means local market which is defined as accessibility of producers directly to market and section of consumers. It provides the most attractive opportunity to the small farm producers looking to large competitive market against the scale economies. In the food sector, organically produced food grains, fruits and vegetables are attractive product choices to a small group of consumers. Still the choice to pursue a niche marketing strategy will not guarantee success, and like any other business decision requires critical assessment.

Niche markets consist of a group of consumers who represent a small segment of the whole market demand, and have similar demographic and buying behaviour. In case of organic products, there are few buyers who prefer quality assurances and source of production while others do not. Even consumers with the same buying behaviour may have different motivations to buy a specific product. For example, organic consumers used to be identified by their concern for their environment, but now there are entirely different segments of consumers who buy organic products because of their perceptions of the higher nutritional benefits from those foods. This chapter explores theoretical explanation of niche market behaviour in case of organic input and output market, and impact of fiscal policy on equilibrium determination.

PRODUCER BEHAVIOUR AND NICHE MARKET

The purpose of the worldwide organic movement is to increase consumers' consciousness on conventional food and its impact on the health of human and environmental domains. Though some agriculture economists express their doubts on the productive efficiency of organic farming, some others do show their interest in it. The above paradox makes it compulsory to study the behaviour and the conditions of the organic producers worldwide (Park & Lohr, 1996). Most of the empirical studies that deal with the producer behaviour mainly focus on farmers and farms characteristics; studies have found that the farms characteristics matter in the decision to adopt organic farming (Burton et al., 2003; Wheeler, 2008; Wynen, 1990). Others who deal with the yield comparison have concluded that the organic farming has lower yields than the conventional (Kleffer et al., 1977; Lansink et al., 2002; Mayen et al., 2010; Nieberg & Offerman, 2000). Only few researches have been carried on the policy instruments which are able to enhance the emergence of organic farming (Dimitri & Oberholtzer, 2005; Eerola & Huhtala, 2008; Jaeck et al., 2012). Thus the growth of organic sub-sector is only possible if the prices of the organic products are higher than the common conventional products or if any policy provision which compensates the productivity gap (Mayen et al., 2010). Although the higher willingness to pay for organic products is established by the various studies, namely Boccaletti and Nardella (2000), Gil et al. (2000), Yiridoe et al. (2005), Krystallis and Chryssohoidis (2005) and Batte et al. (2007), the elimination of productivity gap is related to the integration of science and technology.

Theoretical explanation of the farmer's choice between two technologies of agriculture production; conventional and organic farming has been illustrated by Jaeck et al. (2012). They have tried to examine economic conditions of organic farming under imperfect competition market structure. They have proposed a theoretical approach to explain the behaviour of input supplying firms and farmer's technology choice of organic and conventional technologies; the conventional farming under which chemical fertilisers and high yielding variety seeds, two complementary inputs bundles are used, and other organic farming under using only organic seeds. They have made a strong assumption that in case of organic farming only organic seeds are used in production process. But, the organic producer also uses organic manure whether it is homemade or purchased from the market. Thus, organic farmer also uses a bundle of organic seeds and compost, but its cost is relatively low as compared to the conventional farming.

The demand of inputs depends on the cultivation area, demand of product, piece of input, farmer's income and climatic conditions, etc. But in case of organic farming, demand schedule of inputs depends on the certified area, interests of farmers, income, price of organic inputs, prices of chemical fertilisers, etc. It also indirectly depends on the demand of organic products, prices and the marketing channel. The existence of marketing channels and availability of organic inputs influences the sales of organic inputs. On the one hand supply schedule of organic inputs also depends on the certified area, prices of organic inputs and prices of other chemical inputs in the market. The equilibrium of input market is determined at the point of intersection of aggregate demand and aggregate supply schedule. For theoretical explanation of input equilibrium, we followed Jaeck et al. (2012) type model with minor changes that organic farmers also applied a bundle of organic seeds and compost for increasing their production. The market structure of upstream input firms is oligopolistic in nature and followed Cournot type behaviour for selling their inputs. The equilibrium condition is obtained by the free entry conditions in both sectors is obtained when no producer change their modes of production. It could also happen that multiple equilibriums exist, with or without the appearance of organic sector, to explain viability conditions for organic farming. Subsequently, the incorporation of an 'infant industry' argument is appropriate for policy mechanism, which may help to promote organic farming in developing countries.

As Jaeck et al.'s (2012) theoretical model included two basic approaches which are closely associated with the organic farming: one is 'learning-by-doing' approach and another is 'niche market' approach. The former concept is borrowed from the work of Sipiläinen and Oude Lansink (2005) Martini et al. (2004) and Hanson et al. (1997), which was based on the thought that the conversion of organic farming requires some specific knowledge or begins by some rich and educated farmers. As Sipiläinen and Lansink (2005) estimated technical efficiency of organic farming for furnished dairy farms and found that the average efficiency decreases first at a decreasing rate and then after completing conversion period, it turns toward increasing yield due to 'learning by doing effects'.

The consumers' willingness to pay higher price for organic products is established by various studies (Batte et al., 2007; Boccaletti & Nardella, 2000; Gil et al., 2000; Krystallis & Chryssohoidis, 2005; Yiridoe et al., 2005), which means the richer and urban people's demand for organic products. Thus, organic producer cultivated organic foods for both local urban niche and the worldwide market. But small and marginal farmers are more concerned with local markets, where they can sell their product directly to the consumers. But unfortunately, the 'learning by doing effect' attracts new farmers and decreases profit opportunity with increasing number of organic producers. However, decisions related to choosing a particular farming method, are influenced by farm constraints as well as farmers' preferences (Jaeck & Lifran, 2009). Along with the above adoption of new technology, organic farming also depends on the market structure. It establishes relationships with both upstream and downstream input and output sellers. The existing literature emphasises the oligopolistic market structure of agriculture food under which price transmission and profit are realised by the upstream input sellers (Jack et al., 2012; McCorriston et al., 1998; Rogers & Sexton, 1994; Saitone et al., 2008; Weldegebriel, 2004).

The behaviour of upstream input seller is recognised as an oligopoly market firm behaviour (Fulton & Giannakas, 2001; Hayenga, 1998). In this type of market structure, the strategic behaviour of upstream firms is induced to increase their sale through merging (Johnson & Melkonyan, 2003; Shi, 2009). Furthermore, Just & Hueth (1993) estimated that the combined supply of both types of inputs bundles by a single seller recorded more than separately supply of two input bundles.

In the theoretical explanation Jaeck et al. (2012) assumed that the agricultural inputs provider sells both types of bundles to both the farmer groups. As a result, these input sellers have a significant influence on the adoption of new farming technology. To explain input provider behaviour, they presented a two agricultural inputs bundle supplier model: where chemicals and seeds are assumed complementary and jointly sold by upstream input providers. In the conventional sector, input provider supplied 'bundle of two inputs' chemical fertilisers and high yielding seeds, while for the organic sector; only organic seeds are supplied. But only organic seeds without manure cannot explain the production function. The suppliers of both types' inputs may be different in developing countries and in some cases supplied both types of inputs.

In the imperfect competition market, how the farmer and upstream market behave that helps in emergence and viability of organic farming. Jaeck et al. (2012) suggested a three-step game, in the *first* step producer choose their mode of production; either conventional or organic. The choice of specific farming technology is based on the higher expected return from each type of farming method and the potential effect of learning by doing process. Furthermore, under free entry–exit condition in both sectors, we reached an equilibrium condition where no producer wants to change his/her farming method. In the initial stage, this type of equilibrium condition provides some hope for emergence of organic sector.

In the *second* step, a case when joint input providers choose the quantity of both types of input bundles and sell them to two input markets. Another case when there are two different markets, organic and conventional, each has a small number of input sellers and sell their inputs in their respective markets. Then the competition occurs within the organic and conventional input market firms or suppliers. In the case of a joint input supplier to both groups farmers, Cournot type equilibrium is applied under which these input providers receive the profit and capture both organic and conventional sectors. But in the case of two separate markets for two sectors, organic and conventional firms face dual competition, competition within the industry and with the other rival industry which are producing and selling organic fertilisers and seeds to their own market. The plan and scheme of selling their own inputs influence the farmers to shift production technology.

In the final *third* step organic farming has emerged as the leading sector and the organic products are either sold in an 'urban niche market' or the global conventional products market. Then the organic farmers have two options to sell their products and solve game under Nash equilibrium strategy. On the other hand, when there is another worldwide market of organic product, the organic farmers have three options; either to sell in 'niche market' or in worldwide conventional market or worldwide organic market through national government agency. It is clear that organic farmers have more options to sell their products than conventional growers. By solving this game, farmers attain gain and are closer to the appearance of new organic market and the development of the new farming technology. Consequently, it is essential to analyse a set of policy instruments that can help policy makers to promote organic farming. It means that the growth of organic industry can be boosted by policy support; the argument to support this 'infant industry' is to protect it from the upstream oligopoly competitive market. The competition-oriented policy can promote social welfare arguments that legitimise the support of 'environmental friendly technologies' (Eerola & Huhtala, 2008).

However, imperfect market structure has some specific constraints that need to always keep in mind the policy design. It has assumed that the Government is unable to manage the degree of competition between the upstream inputs providers. It can be used only through conventional policy instruments such as; a lump sum tax on synthetic fertilisers, subsidy to organic seeds, manure and compost and in certification, some specific policy provision for compensation of productivity loss doing conversion period. As the imperfect market structure exists in the agricultural sector, the organic farming can emerge in all conditions and compete with the conventional one.

THE ECONOMIC MODEL FOR NICHE MARKET

For simplification take an economy having two groups of farmers in agricultural sector. One larger group produces a common product devoted to a national and global market called conventional and other small group produces organic products and targets an urban-niche market. Both groups' farmers purchase a bundle of seeds and manure/ chemicals inputs from a small number m of upstream input providers which have some market power. Under this type of input market structure, if the total number farmers are N and each farmer is free to choose either organic or conventional farming, and further suppose the total number of organic farmers is denoted by n, hence the number of conventional farmers will be N-n. This farmer classification and market structure model is borrowed from Jack et al. (2012).

In the conventional sector, chemicals fertilisers and seeds are treated as complementary inputs, similarly for organic sector, seeds and manure are complimentary. For simplification it is assumed that the upstream firm typically sells a bundle of fixed proportion of chemicals and seeds at price p_{bc}, as Hayenga (1998) explains a link between seed and chemical fertilizer market and has concluded that the selling policy of input suppliers is 'to tie the seed customer more closely to the chemical product'. For simplification, we further assume that the farm size is given, and the farm household allocates his/her working time in agricultural related work and leisure. In real life, farmers are constrained by water and input availability, but it has been assumed that farm household in not constrained by water or other inputs availability. This simplification assumption reduces the number of inputs in the production function: the amount of conventional high yielding seeds is denoted by s_c. In function farm denoted as $f(s_c)$ and showing both increasing and decreasing return to scale, that is, $f'(s) > 0$ and $f''(s) < 0$, and also satisfied the *Inada conditions*, that is, $\lim_{s \to 0} f'(s) = +\infty$, $\lim_{s \to \infty} f'(s) = 0$ and does not allow 'free lunch' to anyone, that is, $f(0) = 0$. In this model they have initiate two additional assumptions: one is the elasticity $e_{f''}(s)$ of f'' remains bounded and another is the elasticity $e_{f''}(s)$ of f'' is larger than -2.[1] The output of common product under conventional farming is sold in a worldwide competitive market, at a given market at price p_c. This simplified axiom offers to treat the conventional sector as pure competitive and to primarily concentrate of relation between producer and input providers.

---

[1] Vertical type structure has interlinked marginal productivity (MP) and the demand of inputs. In optimisation problem of input supplier, these restrictions help in first and the second order conditions and rarely meet iso-elastic production function norms.

While organic sector under assumed conditions, it does not employ bundle of synthetic fertilisers and high yielding seeds but apply a bundle of organic manure and seeds at price p_{bo} in place of chemical fertiliser and seeds bundle. Here our model makes a difference with Jack at al. (2012). They had assumed that upstream firms sell only organic seeds and not organic manure but during empirical enquiry of organic farming in Gujarat, India, it has been found that farmers also purchase organic manure along with seeds. Generally, farm households prepare organic manure in their farms and sometimes also purchase from upstream input supplier firms. The production function of this sector is similar as conventional sector depends on the amount of organic seeds and manure bundle, replacing with chemicals, pesticides and seeds bundle. The marginal productivity of inputs is reduced by some factors $\beta \in \{0,1\}$, and which is given as $\beta f(s_o)$. Emergence of organic sector is controlled through two balancing effects, one is 'learning-by-doing effect', and another is 'niche market effect', as per the views of Rouvière & Soubeyran (2011) and Jack et al. (2012).

The 'learning-by-doing effect' encourages farmers to join organic farming and increasing n organic producers, the productivity gap between both sectors falls. It means, that $\beta(n)$ is increasing with the increase of n number of adopters, the productivity gap may disappear when farmers join organic farming. Furthermore, it is logical to say that marginal product of a new adopter is first decreasing with the increasing number producer in organic sector. Additionally, it is also assumed that $\beta'(n) > 0$, $\beta''(n) < 0$ and $\lim_{n \to N} \beta'(n) = 0$.[2] However, the $\beta(N) < 1$: shows that organic sector remains less productive than conventional sector. So, it is justified that the learning by-doing can help in the emergence of organic sector.

Another instrument is 'niche market effect' which implies that health consensus of rich consumers is ready to pay more prices for organic food items. When, the model is more concentrated towards supply side, it does not openly allow this price discrimination behaviour. Here, assumed that the price $p(n)$ of organic food products

---

[2] For simplicity, n is considered as a continuous variable.

depends on the number of organic adopters, at least for the first entrant, attractive enough, that is, $p(0) > p_c$. Though, they also assume that the potential benefit decreases with increasing number of organic growers, even at an increasing rate. The above assumption requires $p'(n) < 0$, $p''(n) < 0$ and that $\lim_{n \to 0} p'(n) = 0$.

As Fulton and Giannakas (2001) and Jaeck et al. (2012) initiated that upstream input suppliers, indexed by $j = 1, 2, 3, \ldots, m$ and assumed it having significant market power. Here two input markets are distinguished: one for the organic producer bundle of manure and seeds, and another for conventional grower's bundles of chemical fertilisers and seeds. In conventional farming these two inputs seeds and synthetic fertilisers, are used in a fixed proportion. Each supplier sales both types of inputs bundles having a fixed proportion, which is denoted by s^o_j and s^c_j the amount of organic manure and seeds bundle and the bundle of seeds and chemicals by seller j. Finally, it assumed that both organic and non-organic inputs are produced at a constant marginal cost, which is denoted by c_o and c_b respectively. Furthermore, it assumed that $c_o < c_b$, which means that the cost of production of organic inputs bundle is lower than the conventional inputs bundle. Thus, the organic producers have competitive advantage, but the input suppliers do not sell inputs at the marginal cost: they always try to capture a part of the producer's profits.

In this game, Cournot type behaviour is quite usual. In free entry condition, farmer chooses as a *first step* whether he/she wants to produce organic or common conventional products. Since entry is free, this choice of method is motivated by the expected gains from each farming method. In the *second step*, the upstream input suppliers choose their optimal supply of both types of input bundles and expected impact of their strategies on both types of producers. In the *third step*, both organic and conventional producers choose optimal quantity of bundles in competitive market structure. So, the equilibrium, under this sub-game, allows to examine the emergence of an organic farming and to design the public policy in such a way that promotes organic farming. The economic model is based on following assumptions.

Assumptions:

1. The farm producers are homogenous in all aspects.
2. The producers are free for exit-and entry in any method of cultivation.
3. The producers are producing under oligopolistic input market structure.
4. The existence of urban niche market effect.
5. The 'learning by doing' effect partially compensates the knowledge gap between organic and conventional farming technologies.
6. The perfect competition in market structure for selling conventional products.

EQUILIBRIUM IN INPUTS SECTOR

Consider the upstream input suppliers who sell input bundles to both producers groups. However, the distribution of input bundles in both the sectors is crucial to know the suppliers behaviour and marketing strategies. So, here our focus is step two and three games, which provide a mechanism to compute the profit of both types of producers and predict emergence of organic farming. Let's, first consider the competitive market structure, where both types of producers have a profit maximisation goal, under which each buy a bundle of inputs until the marginal productivity of input is equal to its market price. In functional form it can be written as:

$$\left\{ \begin{array}{c} p(n).(\beta(n).f'(s_o)) = p_{bo} \\ p_c.f'(s_c) = p_{bc} \end{array} \right\} \tag{7.1}$$

Furthermore, it is keeping in mind that all producers (organic and conventional) are symmetric within the sector. Thus, the inverse demand functions can be obtained for the both types of farmers in the following form;

$$P_{bo}(S_o, k\,(n)) = k\,(n) \cdot f'\left(\frac{S_o}{n}\right) \tag{7.2}$$

$$P_{bc}(S_c, p_c) = p_c \cdot f'\left(\frac{S_c}{N-n}\right) \qquad (7.3)$$

Where S_o and S_c are the aggregated demand of organic and conventional input bundles, expressed as $k(n): = p(n) \cdot \beta(n)$.

In the Cournot type game, the input suppliers choose the optimal supply of both types of bundles s^j_o and s^j_c in such a way that their profit attains from business will be maximum. It means in this type game Nash equilibrium can be shown below with given: $\forall_j = 1, \ldots, m$

$$(\bar{s}^j_o, \bar{s}^j_c) \in \text{argmax}_{(\bar{s}^j_o, \bar{s}^j_c)} P_{bo}\left(\sum_{j=1}^{m} s^j_o, k(n) - c_o\right) \cdot s^j_o$$

$$+ P_{bc}\left(\sum_{j=1}^{m} s^j_c, k(n) - c_b\right) \cdot s^j_c \qquad (7.4)$$

The equation (8.4) shows below as the first order condition of game:

$$\forall_j = 1, \ldots, m \begin{cases} k(n) \cdot f''\left(\frac{1}{n}\sum_{j=1}^{m} s^j_o\right) \cdot \frac{s^j_o}{n} + \left(k(n) \cdot f'\left(\frac{1}{n}\sum_{j=1}^{m} s^j_o\right) - c_o\right) = 0 \\[3mm] p_c \cdot f''\left(\frac{1}{N-n}\sum_{j=1}^{m} s^j_c\right) \cdot \frac{s^j_c}{N-n} + \left(p_c \cdot f'\left(\frac{1}{N-n}\sum_{j=1}^{m} s^j_c\right) - c_b\right) = 0 \end{cases}$$

$$(7.5)$$

Moreover, under the technical assumption, the elasticity $e_{f''}(s)$ of f'' is > -2, which is necessary and sufficient for optimality.

In case of market clearance, a Cournot type equilibrium exists and if the producers of each sector is symmetric, then it can be stated that $\frac{1}{n}\sum_{j=1}^{m} s^j_o$ and $\frac{1}{N-n}\sum_{j=1}^{m} s^j_c$ are, respectively, the amount of seeds and manure/chemical bundles, s_o and s_c applied via organic and conventional producer in the Cournot type equilibrium. It can be observed from equation (7.5) that sales are nearly the same for each input supplier at Cournot type equilibrium, which is specified as:

$$\forall_j = 1,\ldots,m\,(s_{o,}^j\,s_{c,}^j) =$$

$$\left(n\cdot\left(\frac{f'(s_o)-(\frac{c_o}{k(n)})}{-f''(s_o)}\right),(N-n)\cdot\left(\frac{f'(s_c)-(\frac{c_b}{p_c})}{-f''(s_c)}\right)\right) \quad (7.6)$$

After adding the quantities of all types of input providers along with applying all the market clearing restrictions, in Cournot type equilibrium in input market, the game solving problem is obtained for (s_o, s_c) in the following way:

$$\left\{\begin{array}{l}\dfrac{1}{m}\cdot f''(s_o)\cdot s_o + f'\cdot(s_o) = \dfrac{c_o}{k(n)} \\[3mm] \dfrac{1}{m}\cdot f''(s_c)\cdot s_c + f'\cdot(s_c) = \dfrac{c_b}{p_c}\end{array}\right\} \quad (7.7)$$

Under our assumptions, the conclusion follows:

Lemma 1: *This method has a unique solution for (s_o, s_c) and presents unique equilibrium of input supplier in the Cournot type game.*

Although the above lemma 1 is technical but it is necessary for the unique solution. It allowed the input diversification bundles using different quantities and packaging as selling strategy. In this type of equilibrium both types of producers and even the input suppliers make the profit, by optimum purchase and by optimum sale of input bundles, respectively.

In equation (7.7), $s_o\left(\dfrac{c_o}{k(n)}, m\right)$ and $s_c\left(\dfrac{c_b}{p_c}, m\right)$ shows the equilibrium demand of input bundles made by both types of producers. These two functions show the relative quantity of both the types input bundles used by both the sectors or sale m number of input suppliers and the ratio of the cost over the price, which is relative profit of each sector.

Here, also allowing the effect of learning-by-doing process on adaptation, we observe the following:

Proposition 1: the equilibrium demand of organic input bundles,

$s_o\left(\dfrac{c_o}{k(n)},m\right)$ in organic sector and $s_c\left(\dfrac{c_b}{p_c},m\right)$ conventional input

bundle in conventional sector used by the respective producer is

decreasing with the ratio $\dfrac{c_o}{k(n)}$ and $\dfrac{c_b}{p_c}$, and increasing with the degree

of competition measured by the m number of input suppliers. However as $m \to \infty$, these quantities of input bundle sales reaches toward the

competitive equilibrium quantities $s_o\left(\dfrac{c_o}{k(n)}\right) = (f')^{-1}\left(\dfrac{c_o}{k(n)}\right) = $ and s_c

$\left(\dfrac{c_b}{p_c}\right) = (f')^{-1}\left(\dfrac{c_b}{p_c}\right) = $ respectively.

It is assumed that each producer's behaviour is competitive and adjusting their marginal gain obtained from the input cost to its price from equation (8.1), it is easy to calculate the gains of each type of producer. The profit functions of both types of producers are written as:

$$\left\{ \begin{array}{l} \pi_o\left(k(n),c_o,m\right) = \\[4pt] k.\left[f(s) - f'(s)\cdot s\big|_{s_o\left(\frac{c_o}{k(n)},m\right)} \right] = k(n)\cdot\gamma(s)\big|_{s_o\left(\frac{c_o}{k(n)},m\right)} \\[10pt] \pi_o(p_c,c_b,m) \\[4pt] = k.\left[f(s) - f'(s)\cdot s\big|_{s_c\left(\frac{c_b}{p_c},m\right)} \right] = p_c\cdot\gamma(s)\big|_{s_c\left(\frac{c_b}{p_c},m\right)} \end{array} \right\} \quad (7.8)$$

From equation (7.8) it is observed that profits are non-negative since for all neoclassical production functions $f(s)$ the marginal productivity is always lower than the average productivity[3] so that $\gamma(s): = f(s) - f'(s)\cdot s \geq 0$.

---

[3] It shows the absence of 'free lunch' (i.e., *f (0) <hig>=</hig> 0*) and the concavity off.

In the same stratum, it can be computed from lemma 1 using the quantities of organic and conventional input bundles, sold by each input supplier. By rearranging equation (7.6) the quantities of both input bundles, we find:

$$
\left\{
\begin{aligned}
s_o^j\left(\frac{c_o}{k(n)},m,n\right) &= \frac{n}{m} \cdot s_o\left(\frac{c_o}{k(n)},m\right) \\
s_{c,}^j\left(\frac{c_b}{p_c},m,n\right) &= \frac{N-n}{m} \cdot s_c\left(\frac{c_b}{p_c},m\right)
\end{aligned}
\right\}
\tag{7.9}
$$

And the profit of each input provider firm can be expressed by the following functional farm:

$$
\pi\left(k(n),c_o,p_c,c_b,m,n\right)=
$$

$$
\frac{1}{m^2}\left(
\begin{aligned}
&n\cdot k(n)\cdot\left[-f''\left(s_o\left(\frac{c_o}{k(n)},m\right)\right)\right]\cdot\left(s_o\left(\frac{c_o}{k(n)},m\right)\right)^2 \\
&+(N-n)\cdot p_c\cdot\left[-f''\left(s_c\left(\frac{c_b}{p_c},m\right)\right)\right]\cdot\left(s_c\left(\frac{c_b}{p_c},m\right)\right)^2
\end{aligned}
\right)
\tag{7.10}
$$

These results have some logic, although some scholars, namely Saitone et al. (2008), have observed that the selling behaviour of upstream suppliers in the input sector has vital distributional impacts under imperfect competition. The upstream input supplier market power is measured via the inverse of the m number of input suppliers in the market. As the number increases, typically the distribution of input amount reduces in both the sectors. This condition is also shown in lemma 2 within the competitive market structure. This results in change in profit distribution of input providers because their share reduces with increase in the number of distribution. Thus, the profit of input provider reduces to zero, at perfect market competitive structure, with constant returns to scale. Thus, the profit of input suppliers disappears when the number of sellers becomes very large. In such a situation, input sale reduces, and the resultant profit of the input providers reaches zero.

The Case of Free Entry Condition

In case of free entry or exit in the market, farmers have freedom of choice regarding which farming method to adopt. But choice of a farming method depends on the net returns or profit. The emergence of organic farming is possible if profit in organic farming is greater than conventional agriculture.

Equilibrium Distribution in Case of Free Entry

It is essential to define the concept of equilibrium distribution with free entry–exit condition. The equilibrium distribution in both the sectors is attained, when no farm producer expects higher return or not willing to shift in other sector due to zero gain. The free entry means that farm producer can shift one sector to the other without restriction. The profit of the conventional producers is independent of the number of organic producers as shown in equation (8.8). It means the equilibrium distribution is attained at n^* with this property shown as:

$$\left\{ \begin{array}{l} \pi_o\left(k(n^*)c_o, m\right) \geq \pi_c(p_c, c_b, m) \\ \pi_c(p_c, c_b, m) \geq \pi_o\left(k(n^*-1)c_o, m\right) \end{array} \right\} \quad (7.11)$$

Here, no organic grower is willing to shift to the conventional method. Similarly, in conventional sector no conventional producer is willing to shift to the organic sector. For simplification, if we consider that n is a continuous variable and satisfying these two basic properties:

$$\left\{ \begin{array}{l} \pi_o\left(k(n^*)c_o, m\right) = \pi_c\left(p_c, c_b, m\right) \\ \pi_o\left(k(n^*)c_o, m\right) \text{ is decreasing at} n^* \end{array} \right\} \quad (7.12)$$

Further inquiry of the equilibrium distribution is important to find out the profit motive behaviour of the organic grower when n varies. Taking the partial derivative with respect to n, the result is:

$$\frac{\partial \pi_o}{\partial n} = k'(n) \cdot \left[\left(\gamma(s) \big|_{s_o\left(\frac{c_o}{k(n)}, m\right)} \right) - \left(\frac{d\gamma}{ds} \big|_{s_o\left(\frac{c_o}{k(n)}, m\right)} \right) \cdot \frac{\partial s_o}{\partial \left(\frac{c_o}{k(n)}\right)} \cdot \frac{c_o}{k(n)} \right] \quad (7.13)$$

From the economic theory of production, we know that:

- $\gamma(s) \geq 0$ when the marginal productivity < the average productivity,
- via proposition 1, $s_o\left(\frac{c_o}{k(n)}, m\right)$ is declining with $\frac{c_o}{k(n)}$ and,
- via calculation $\frac{d\gamma}{ds} = -f''(s) \cdot s \geq 0$

Here it shows that the sign of $\frac{\partial \pi_o}{\partial n}$ should be similar to the sign of $k'(n)$. In other words, if π_o (k (n), c_o, m) is falling, n is not effectively explained by the interaction of learning by doing process $\beta(n)$. Along with this, the advantage of the 'niche market effect' is measured by $p(n)$. So, the first effect on the profit of organic producers will be decreasing with the increasing number of organic growers, that is, $\beta''(n) < 0$, while the existence of urban niche market may make benefits but it also decreases with the increase in organic producers, that is, (n) < 0 and $p''(n) < 0$, it is expressed as:

Lemma 2: *The profit function of an organic grower is inverted ∩-shaped in n. It, first increases because of the gain obtained from the 'learning by doing effect' and then dominated by the losses induced by the erosion of the price in the 'niche market'. The learning by doing effect works up to a certain number of n_{max}, while the erosion of the price will dominate after that number.*

This inverted ∩-shape of profit function has many interpretations in the appearance of organic farming within agriculture sector. First, at the certain n_{max} number, the profit of organic producers is less than the returns received by conventional producer, that is,

$$\pi_o(k(n_{max}), c_o, m) < \pi_c(p_c, c_b, m) \quad (7.14)$$

In this case, organic sector farming will never emerge. This situation occurs only due to the maximum gains earned via 'learning by doing effect' is unable to compensate the productivity loss of the organic producers. If the equation (7.14) condition is not met, then one can hope that the organic farming occurs for sure. Since the inverted ∩-shaped property has some interpretations it may occur that the profit of the first producer who adopted organic farming receives less than that obtained by the conventional producer, that is,

$$\pi_o(k(1), c_o, m) < \pi_c(p_c, c_b, m) \qquad (7.15)$$

In this situation, two types of equilibrium distributions emerge:

1. One situation, if both conventional and organic farming or merely organic farming, if $\pi_o \{k (N), c_o, m\} > \pi_c (p_c, c_b, m)$.
2. Another situation, when only conventional farming emerges, it is simply due to first adopter of organic farming, receives the benefits from 'learning by doing effect' and does not get additional gain from 'niche market effect' he may not able to compensate the productivity losses.
3. Finally, if both equations (7.14) and (7.15) are not met then it can expect a unique equilibrium distribution in both sectors producers and only if $\pi_o \{k (N), c_o, m\} > \pi_c (p_c, c_b, m)$.

The Emergence of Organic Farming under Imperfect Competition

Now let's move a step forward and try to understand the emergence of organic farming and identify what are the favourable conditions. Now, we consider calculating jointly the cost of both type of input bundles and the prices for conventional and organic products along with the 'learning by doing effect' and 'niche market effect'.

Let's, first consider the simple problem, where the organic farming sector does not have any competitive advantage, that is,

$$\forall_n, \frac{c_o}{p(n)\beta(n)} > \frac{c_b}{p_c} \qquad (7.16)$$

As discussed above, organic farming never emerges if the new adopter has no gain. A mechanism can be built up in this situation that is different from competitive structure and is driven by the upstream imperfectly competitive input providers. Though these input providers know about the advantage of competitive market, but they have a quantity distribution input strategy that favours conventional producers. It can be inferred directly from equation (7.7), (7.2) and (7.3): under given conditions (7.16) and for any distribution of n number producers between both the sectors. So, they are willing to sell less organic input bundles at a higher price to organic producers. The use of the above restriction reduces the profit opportunity of upstream input providers from organic farming. Moreover, since it is assumed that $c_0 < c_b$, equation (7.16) shows that $\forall_n$, $p(n)\,\beta(n) < p_c$, that is, the joint effect of the 'learning-by-doing' and the 'niche market' is unable to attain the price of conventional products. The disadvantage of this condition is that the output market discourages conventional producers.

This conflicting thought is tried to make sense by assuming that if the organic producers have at least such a distribution n, where they have competitive gains, then there is only potential to emerge as an alternative farming sector. This perception is because of assuming imperfect competition market structure; however, it is again incorrect. When consider, the above case, where the upstream inputs sellers support the organic producer through selling more organic input bundles at a lower price. Then they are able to capture a large share of their profits as $(ef''\,(s) > -2)$. From the above point of view this general setting,[4] leads to generate confidence that organic producer can also sell their organic products at a higher premium price in the 'niche market' and also receive benefit from the 'learning-by-doing effect', that is, $\exists_n$, $p(n)\,\beta(n) \geq p_c$. it can be also expressed said as:

Proposition 2: Let $\max_n$ is taken as describe in lemma 2. Since it's focused on the appearance of organic farming, expressed as:

---

[4] When we use a general production function with a given sufficient condition, most of the parameters are calculated using constant elasticity production function.

1. If $\max_n p(n)\beta(n) \in \left[0, \dfrac{c_0}{c_b}, p_c\right[$ organic farming never occurs,

2. If $\max_n p(n)\beta(n) \in]p_c, \infty$ [an equilibrium distribution exists for organic producer,

3. If both the above conditions are satisfied, organic farming emerges only if

$$\pi_o\left(k(n_{max}), c_o, n_{max}, m\right) < \pi_c\left(p_c, c_b, m\right)$$

Furthermore, even if an equilibrium distribution exists for organic producers under free entry conditions with (2) and (3), the other equilibrium distribution may appear only conventional producers since the first organic adopter has no gains from organic farming. This case is verified as $\pi_o\left(k(1), c_o, m\right) < \pi_c\left(\dfrac{c_b}{p_c}, m\right)$.

The final remark on the ∩-shape of profit function for organic producer is that it is linked with the facts of market conditions and cost of production. If the final condition is not fulfilled, the growth of organic sector can suffer from the problem of coordination. So, if the Government wants to support organic sector, then there is possibility of emergence of organic farming. But, before dealing the policy issues, we need to explain the major properties of the equilibrium.

EQUILIBRIUM IN ORGANIC SECTOR

Since our prime interest is organic sector growth, we have assumed;

$$\pi_o\left(k(n_{max}), c_o, n_{max}, m\right) < \pi_c\left(p_c, c_b, m\right)$$

In order to ensure organic sector equilibrium, we need to understand three major equations;

$$\left\{\begin{array}{l} \varnothing(s_o) = \dfrac{1}{m}\cdot f''(s_o)\cdot s_o + f'\cdot(s_o) = \dfrac{c_o}{p(n)\cdot\beta(n)} \setminus \\[3mm] \varnothing(s_c) = \dfrac{1}{m}\cdot f''(s_c)\cdot s_c + f'\cdot(s_c) = \dfrac{c_b}{p_c} \quad \text{With}\, \gamma(sc) = f(s) - f'(s)s \\[3mm] p(n)\cdot\beta(n)\cdot\gamma(s_o) = p_c\cdot\gamma(s_c) \end{array}\right\}$$

Here the first two equations show the optimal distribution preferences of input suppliers as per equation (7.7), whereas the third is for the entry–exit condition. Thus, here also φ is falling while γ is rising.

Above functional structure provides a foundation to emphasise the properties of organic sector equilibrium. It is assumed that the profits of both sector producer under free entry condition. But the entry–exit condition also provides a chance to go a step ahead in the explanation of equilibrium condition. It is important to remember here that the assumption, per unit cost of cultivation for organic product using organic input bundle is lower than the cost of conventional input bundle that is $c_o < c_b$. So, the assumption is that the prices of both types' products are equal, and the productivity loss of organic producer is deflated via $pc = p(n) \cdot \beta(n)$. From this we can infer that the optimal strategy of input supplier is selling more organic input bundles to the organic producers simply because $c_o \, p(n) \cdot \beta(n) < c_b$ p_c and $\varphi(s)$ is descaling. But when added to the condition that $\gamma(s)$ is rising and the equilibrium at restricted free-entry condition, then the gains of the organic producers receives more than conventional producers. It encourages the new entrants in the organic sector and contributes to decline of $p(n)$. $\beta(n)$. Therefore, this argument that at equilibrium point the price of organic products is deflated by the losses of productivity, that is $k(n) = p(n) \cdot \beta(n)$ is always less than the price of conventional products p_c. This result supports of free entry–exit condition with the fact that $\gamma(s)$ is decaling and all organic producers apply more seed and manure bundle than conventional producer at equilibrium and on a same plot of land, that is $s_o \geq s_C$. Hence further stated as:

Proposition 3: with a free entry–exit equilibrium condition inorganic sector, it is observed as:

1. the productivity losses are compensated by the higher premium price of organic produce, which is lower than the price of conventional products that is $p(n^*) \cdot \beta(n^*) \leq p_C$;
2. Organic practice requires additional organic input bundles per plots of land than conventional, that is, $s_o^* \geq s^*$.

Public Policy and Emergence of Organic Farming

The key argument of supporters of organic farming is that it is a new rising 'infant industry' within agriculture sector. Therefore, it is needed to be supported and protected by government through policy instruments. The role of public policy is crucial in the development of organic sector, especially in the cost of certification and inspection, input subsidy, market facilities, etc. The government's policies follow the social welfare criteria to make policy announcements. But it would require information about both the consumers as well as producers; the former was not included in the model. However, under imperfect competition structure some policy instruments may be inefficient or have 'perverse effects'. Here discussion has two steps processor: one presents the main instruments and their basic effects, and another has impact on the agricultural sector.

The Set of Entrant Instruments

Initially it might be happened that the joint gain attained through the 'learning-by-doing effect' and the declining gain 'niche market effect' achieved through higher premium prices reaches equilibrium. The circumstances may take place as discussed above, where the first adopter of organic farming attaining lower profit that the conventional producer. This situation is recorded as coordination problems which prevents the growth of organic farming. Government can play a significant role by making policy to overcome this problem by means of increasing the information, tanning services and movement of farming communities. The best idea is to reduce the probable loss of the earliest adopter which encourages others to join organic farming by receiving greater benefits from 'learning-by-doing effect'. We must allow the efficiency of the collusion policy that is inversely related with the size coalition that realises a better profit in organic farming, that is, having lesser number n_{min} of agents;

$$\pi_o(k(n_{max}), c_o, n_{max}, m) < \pi_c(p_c, c_b, m) \qquad (7.17)$$

If an association composed of n_{min} agents adopts organic cultivation with natural entry market equilibrium condition, this argument indirectly supports a policy of 'infant organic industry'.

When this type equilibrium condition occurs, a government policy may help organic farming via appropriate fiscal policy tools. For example; tax on the chemicals and pesticides, subsidy to organic producers on organic input bundles, subsidy for certification cost and a subsidy for productivity losses are standard fiscal policy tools, etc. These fiscal policy tools are applied in the model:

1. A fixed tax on per unit conventional input bundles and added with cost of cultivation c_b with a rate τ of per bundle (τ is a proportion of pesticide and chemicals in the bundle), tax burden falls indirectly on input seller.

2. A lump sum subsidy σ per unit of organic seed and manure bundle given to organic producer reduces their cultivation cost measured by p_s the price of organic seed and manure bundle as per equation (7.1). From equations (7.2) and (7.4), it decreases the inverse demand function by σ. As a result, it is equal to reduction of unit cost of cultivation by subsidy that is $c_o - \sigma$.

3. A subsidy δ per unit on organic product, it delivers an additional gain for organic producer and in the price $p(n)$ in equation (7.2).

4. The impact of adjust in the 'learning by doing curve' is more difficult to capture. it is basically stated that curve shifts up by some λ so we add λ in equation $\beta(n)$ by $\beta(n) + \lambda$.

Making rearrangement of equations (7.7) and (7.12) having assumption that a new equilibrium exists in the regional interior equilibrium distribution (i.e., $n^* < N$). So, the outcomes of these fiscal policy tools are deliberating via using the implicit function theorem:

$$\left\{ \begin{aligned} \varnothing(s_o) &= \frac{1}{m} \cdot f''(s_o) \cdot s_o + f' \cdot (s_o) = \frac{c_o - \sigma}{k(n)} \\ \varnothing'(s_c) &= \frac{1}{m} \cdot f''(s_c) \cdot s_c + f' \cdot (s_c) = \frac{c_b + \tau}{p_c} \\ k(n) \cdot \beta(n) \cdot \gamma(s_o) &= p_c \cdot \gamma(s_c) \end{aligned} \right\}$$

With $k(n, \delta, \lambda) = (p(n) + \delta)(\beta(n) + \lambda)$

(7.18)

Proposition 4: If we consider equilibrium in the regional distribution of both types of producers, then it becomes easy to sum up the effect of the fiscal policy tools on the production level of both sectors producers and on the equilibrium distribution. The results are presented in Table 7.1.

Effects of Fiscal Policy Tools

Let us begin with a tax on chemicals and pesticides bundles on conventional producers. Taxation is a fiscal policy tool and has 'indirect and contrasted effect' on organic producers. A tax on synthetic fertilisers raises the price conventional input bundles as in equation (7.1)

$$\frac{\partial p_b}{\partial \tau} = \frac{\partial}{\partial \tau}\left(p_c \cdot f'(s_c)\right) = p_c \cdot f''(s_c) \cdot \frac{\partial s_c}{\partial \tau} > 0$$

Since the non-competitive suppliers are tried to maintain their profit margins by selling both types of input bundles, their strategy may reduce the gains of the conventional producers and encourage them to adopt organic cultivation. At the same time any entry in the organic sector diminishes the expected profit attained from the 'niche market effect'. Once the equilibrium occurs in the organic sector, the 'niche market effect' dominated the potential gains from 'learning-by-doing effect'. So, a tax on conventional bundles increases the number of organic producers ($\partial n/\partial \tau > 0$ in Table 7.1). It also results in decrease per unit profit of producers and the entry–exit stops, only when the profit of both sectors producers becomes equal.

A direct subsidy for organic seed and manure bundle has a strong effect on organic producers. When the subsidy is directly given to the producer, the input seller may also give some incentives to decrease their margin on organic seeds and manure bundle in order to sell more organic input bundles to each producer. The aim of input seller policy is to a get a part of this subsidy. The question arises as to why the price of organic input bundle decreases as shown in equation (7.1):

Table 7.1 *Taxation and Subsidy Policy Instrument and Its Impact*

	Tax on Chemicals ($\partial\tau$)	*Subs. for Chemical Free Seeds (∂_o)*	*Subs. for Organic Products ($\partial\delta$)*	*Learning by Doing ($\partial\lambda$)*
Seeds used by an organic farmer (∂s_o)	$-\dfrac{\gamma'(s_c)\cdot\phi(s_o)}{D\cdot\varphi'(s_c)\cdot\kappa(n,\delta,\lambda)}<0$	$-\dfrac{\gamma(s_o)}{D\cdot\kappa(n,\delta,\lambda)}>0$	0	0
Bundles used by a Conv. farmer (∂s_c)	$\dfrac{1}{\varphi'(s_c)\cdot p_c}<0$	0	0	0
Distribution of the farmers (∂_n)	$-\dfrac{\gamma'(s_c)\cdot\phi'(s_o)}{D\cdot\varphi'(s_c)\cdot\partial n\kappa(n,\delta,\lambda)}>0$	$-\dfrac{\gamma'(s_o)}{D\cdot\partial n\kappa(n,\delta,\lambda)}>0$	$-\dfrac{\beta(n)+\lambda}{\partial n\kappa(n,\delta,\lambda)}$	$-\dfrac{p(n)+\delta}{\partial n\kappa(n,\delta,\lambda)}>0$

With $D = [\varphi'(so)\gamma(so)-\gamma'(so)\phi(so)]<0$.

$$\frac{\partial p_o}{\partial \sigma} = \frac{\partial}{\partial \tau}\left(\kappa(n,\delta,\lambda)\cdot f'(s_o)\right)$$

$$= \frac{\partial \kappa(n,\delta,\lambda)}{\partial n}\cdot\frac{\partial n}{\partial \sigma} f'(s_o) + \kappa(n,\delta,\lambda)\cdot f''(s_o)$$

$$= \frac{f(s_o)\cdot f''(s_o)}{-D} < 0.$$

Since profit increases, a share of profit held by the organic producers is his saving, as a result this sector attracts new entrants and hence entry takes place ($\partial n/\partial > 0$ in Table 7.1a). The entry will not stop until the additional gains are not eliminated in the sector. So, we can conclude that as organic producers increase in numbers, in the long run they earn the same profit as without subsidies. The important point is that input seller does not change their strategy for conventional sector producers.

$$\frac{\partial s_c}{\partial \sigma} = 0 \text{ and therefore } \frac{\partial p_b}{\partial \tau} = 0$$

When a few conventional producers shifted to the organic sector, it reduces the demand of conventional input bundle. Then input sellers will also receive benefit from the subsidy given to these new organic adopters.

These two fiscal policy tools; one is subsidies for organic producer, which also enrich the 'learning-by-doing effect', have quite identical effects which are mainly explained through the free entry–exist condition under imperfect competitive input market structure. The policy adjustments tools such as: subsidy to organic input and certification or government spend in learning-by-doing process that increases the profit of organic producers because $\kappa(n,\delta,\lambda) = (p(n) + \delta)(\beta(n) + \lambda)$ increases. So, if other things remains unchanged, this encourages entry until $\kappa(n,\delta,\lambda)$ reach a level that equalises profits in both sectors, which is confirmed by:

$$\frac{d\kappa(n,\delta,\lambda)}{d\delta} = \frac{\partial \kappa(n,\delta,\lambda)}{\partial n}\cdot\frac{\partial n}{\partial \delta} + \frac{\partial \kappa(n,\delta,\lambda)}{\partial \delta}$$

$$= \frac{\partial \kappa(n,\delta,\lambda)}{\partial n}\cdot\frac{-(\beta(n)+\lambda)}{\partial n\kappa(n,\delta,\lambda)} + +\beta(n)+\lambda = 0$$

The input suppliers are unable to regulate their profit margins through capturing the additional profit as shown above. The important question here is, 'Why input sellers do not try to attain a part of these subsidies?' To know the answer of above question it is essential to return to the basic equations that specify the behaviour of input sellers as per equation (7.7). It has been observed that these input providers do not adjust their behaviour for conventional seeds and fertiliser bundle result into no change in profit of each producer in this sector. While for organic sector, quite a different strategy is adopted, equation (7.7).

$$\frac{1}{m} \cdot f''(s_o) \cdot s_o + f'(s_o) = \frac{c_o}{\kappa(n, \delta, \lambda)}$$

Hence, if there is restriction on free entry-exist in the market, then any rise of κ (n, δ, λ) leads to increase in the additional subsidies that will increase the quantity demand of seeds and manure bundle sold to each producer, in order to receive a part of the government subsidies. But once free entry–exit condition is allowed, and the input distribution is close to equilibrium with organic sector, Input sellers know that κ (n, δ, λ) also declines, as a result they must decrease incentive. This interaction works until the initial level of sale of organic input bundle is attained. So, in this type behavioural model at free-entry condition, the backward solving game is applied. The fiscal subsidy policy is the most used tool and the public spending goes in the development of organic sector.

CONCLUSION

The theoretical and conceptual model of niche market and producer behaviour is based on the work of Jaeck et al. (2012). The various modification and changes are made to explain the organic niche market input suppliers behaviour and the respect to change policy regime the impact on the both group of producers. The aim of the conceptual model to present a policy formwork which help farmers to adopt organic farming and attain the goal of sustainable agriculture practices. The theoretical model presents different policy alternatives, which have varied outcomes. It depends on the government as to which option one needs to move forward with.

REFERENCES

Batte, M. T., Hooker, N. H., Haab, T. C., & Beaverson, J. (2007). Putting their money where their mouths are: Consumer willingness to pay for multi-ingredient, processed organic food products. *Food Policy*, 32(2), 145–159.

Boccaletti, S., & Nardella, M. (2000). Consumer willingness to pay for pesticide-free fresh fruit and vegetables in Italy. *The International Food and Agribusiness Management Review*, 3(3), 297–310.

Burton, M., Rigby, D., Young, T., (2003) Modelling the adoption of organic horticultural technology in the UK using duration analysis. *The Australian Journal of Agricultural and Resource Economics*, 47 (1), 29–54

Dimitri, C., & Oberholtzer, L. (2005). *Market-led growth vs. government facilitated growth: Development of the U.S. and EU organic argicultural sectors.* WRS-05-05, US Department of Agriculture Economic Research Service/USDA.

Eerola, E., & Huhtala, A. (2008). Voting for environmental policy under income and preference heterogeneity. *American Journal of Agricultural Economics*, 90(1), 256–266.

Fulton, M., & Giannakas, K. (2001). Agricultural biotechnology and industry structure. *AgBioForum*, 4(2): 137151:

Gil, J. M., Gracia, A., & Sanchez, M. (2000). Market segmentation and willingness to pay for organic products in Spain. *The International Food and Agribusiness Management Review*, 3(2), 207–226.

Hanson, J. C., Lichtenberg, E., & Peters, S. E. (1997). Organic versus conventional grain production in the mid-Atlantic: An economic and farming system overview. *American Journal of Alternative Agriculture*, 12(1), 2–9.

Hayenga, M. L. (1998). Cost structures of pork slaughter and processing firms: Behavioral and performance implications. *Review of Agricultural Economics*, 20(2), 574–583.

Jaeck, M., & Lifran, R. (2009, January). Preferences, norms and constraints in farmers' agro-ecological choices. Case study using a choice experiments survey in the Rhone River Delta, France. *2009 Conference (53rd)*, February 11–13, 2009, Cairns, Australia.

Jaeck, M., Lifran, R., & Stahn, H. (2012). *Emergence of organic farming under imperfect competition* (No. 1239). Aix-Marseille School of Economics, Marseille, France.

Johnson, S. R., & Melkonyan, T. A. (2003). Strategic behavior and consolidation in the agricultural biotechnology industry. *American Journal of Agricultural Economics*, 85(1), 216–233.

Just, R. E., & Hueth, D. L. (1993). Multimarket exploitation: The case of biotechnology and chemicals. *American Journal of Agricultural Economics*, 75(4), 936–945.

Klepper, R., Lockeretz, W., Commoner, B., Gertler, M., Fast, S., O'Leary, D., & Blobaum, R. (1977). Economic performance and energy intensiveness on organic and conventional farms in the Corn Belt: A preliminary comparison. *American Journal of Agricultural Economics*, 59(1), 1–12.

Krystallis, A., & Chryssohoidis, G. (2005). Consumers' willingness to pay for organic food: Factors that affect it and variation per organic product type. *British Food Journal*, *107*(5), 320–343.

Lansink, A. O., Pietola, K., & Bäckman, S. (2002). Efficiency and productivity of conventional and organic farms in Finland 1994–1997. *European Review of Agricultural Economics*, *29*(1), 51–65.

Martini, E. A., Buyer, J. S., Bryant, D. C., Hartz, T. K., & Denison, R. F. (2004). Yield increases during the organic transition: improving soil quality or increasing experience? *Field Crops Research*, *86*(2), 255–266.

Mayen, C. D., Balagtas, J. V., & Alexander, C. E. (2010). Technology adoption and technical efficiency: organic and conventional dairy farms in the United States. *American Journal of Agricultural Economics*, *92*(1), 181–195.

McCorriston, S., Morgan, C. W., & Rayner, A. J. (1998). Processing technology, market power and price transmission. *Journal of Agricultural Economics*, *49*(2), 185–201

Nieberg, H., & Offermann, F. (2000). Economic performance of organic farms in Europe. volume 5, Universität Hohenheim, Stuttgart-Hohenheim.

Park, T. A., & Lohr, L. (1996). Supply and demand factors for organic produce. *American Journal of Agricultural Economics*, *78*(3), 647–655.

Rogers, R. T., & Sexton, R. J. (1994). Assessing the importance of oligopsony power in agricultural markets. *American Journal of Agricultural Economics*, *76*(5), 1143–1150.

Saitone, T. L., Sexton, R. J., & Sexton, S. E. (2008). Market power in the corn sector: How does it affect the impacts of the ethanol subsidy? *Journal of Agricultural and Resource Economics*, *33*(2), 169–194.

Shi, G. (2009). Bundling and licensing of genes in agricultural biotechnology. *American Journal of Agricultural Economics*, *91*(1), 264–274.

Sipiläinen, T., & Oude Lansink, A. (2005, August). Learning in organic farming: An application on Finnish dairy farms. In *XIth Congress of the EAAE*, Copenhagen, Denmark.

Weldegebriel, H. T. (2004). Imperfect price transmission: is market power really to blame? *Journal of Agricultural Economics*, *55*(1), 101–114.

Wheeler, S. A. (2008). What influences agricultural professionals' views towards organic agriculture? *Ecological Economics*, *65*(1), 145–154.

Wynen, E. (1990). Sustainable and Conventional Agriculture in South-Eastern Australia: A Comparison. Economics Research Report, no. 90.1. School of Economics and Commerce, La Trobe University, Melbourne.

Wynen, E., & Edwards, G. W. (1990). Towards A Comparison of Chemical-Free and Conventional Farming in Australia. *Australian Journal of Agricultural and Resource Economics*, *34*(1), 39–55.

Yiridoe, E. K., Bonti-Ankomah, S., & Martin, R. C. (2005). Comparison of consumer perceptions and preference toward organic versus conventionally produced foods: a review and update of the literature. *Renewable Agriculture and Food Systems*, *20*(04), 193–205.

Organic Output Market in Asia

INTRODUCTION

The size of input and output market depends on the demand for organic foods, and the demand for organic foods depends on the consumers' behaviour. The consumers' behaviour depends on their income, price of organic products, taste preference, education, environment, awareness and so on. So the market size and structure depend on how many consumers are willing to purchase organic products, which are more expensive than conventional products. What factors will change consumers' consumption behaviour from conventional food basket to organic food basket? The above two questions imply two things; first, how many consumers are ready shift from non-organic to organic food and why? The second question will explain the factors behind this changing behaviour of consumers. These two questions will decide the size and structure of organic input and output. The answers to the first and second questions are related to consumers' buying behaviour and behavioural shift from non-organic to organic food. All countries agree that climate change, need for sustainable development, environmental degradation and health problems are the main factors behind this shift.

In Asia, organic market output and input data is not available for most of the countries, but the most populous countries, China and India, are emerging as producers of organic foods. According to the 2017 Revision, the world's population was nearly 7.6 billion. Another important fact is that 60 per cent of the world's people live in Asia (4.5 billion), 17 per cent in Africa (1.3 billion), 10 per cent in Europe (742 million), 9 per cent in Latin America and the Caribbean

(646 million), and the remaining 6 per cent in North America (361 million) and Oceania (41 million). Both China (1.4 billion) and India (1.3 billion) remain the two most populous countries of the world, comprising 37 per cent of world population (United Nations, 2017).

This chapter provide an explanation of the Asian organic input–output market opportunity, prospects and challenges for the rest of the world. In Asia, India and China are two emerging nations and both have their own opportunities and challenges in this emerging organic market. So Asia will emerge as the leading organic producer if India and China's organic sectors grow faster than other competing nations of other continents. This chapter provide information related to the number of producers, organic output and export–import of these Asian counties with rest of the world.

ORGANIC OUTPUT MARKET IN ASIA

The global market for organic food increased from 7.9 billion dollars in 2000 to 89.7 billion dollars in 2016. All regions had recorded healthy growth rate in 2016, but North America and Europe emerged as twin engines for growth of organic trade. These two regions recorded 90 per cent of organic international trade. United States and Germany reached 5 per cent market share of the world organic market. Other countries of the regions outside of Europe and North America remain below 1 per cent of total world organic product trade (Sahota, 2016).

Table 8.1 presents the share of each region in the global organic market in 2016. The retail sale of organic products was recorded 89.10 per cent combining both North America and Europe. In the global organic market, North America had a share of 49.52 per cent, Europe had 39.58 per cent share and Asia had a share of 8.67 per cent only. The other regions, that is, Africa, Latin America and Oceania jointly had a share which was less than 3 per cent of the global organic market. Thus, Asia is the third largest market of organic products in the world after North America and Europe. North America has the highest per capita consumption (PCC) of organic products in terms of euro currency followed by Europe.

Table 8.1 Global Market of Retail Sale of Organic Products and PCC in 2016

Region	Retail Sale (ml €)	Retail Vale in (%)	PCC (€)
Africa	16	0.02	–
Asia	7,343	8.67	1.7
Europe	33,526	39.58	40.8
Latin America	810	0.96	1.3
North America	41,939	49.52	117.0
Oceania	1,065	1.26	26.5
World	84.698	100	11.3

Source: FiBL statistics (2018).

Table 8.2 shows that the value of organic product exports is continuously increasing in Asia from 17.26 to 2,060.73 million euros during 2005 to 2016, and each year the value of organic export

Table 8.2 Asian Organic Market Value in Euros Million

Year	Export (million €)	Import (million €)	Retail Sale (million €)
2005	17.26	–	38.70
2006	49.18	–	617.59
2007	74.55	–	983.57
2008	381.41	–	1,156.53
2009	437.56	–	1,933.76
2010	470.54	–	1,935.17
2011	480.20	–	1,893.13
2012	780.49	46.58	1,973.97
2013	902.88	46.58	3,613.12
2014	1,520.52	40.7.0	5,068.52
2015	1,848.47	22.95	6,142.18
2016	2,060.73	49.01	7,342.97

Source: FiBL statistics (2018).

Table 8.3 *Growth of Market Size and Operators in Asia*

Year	Organic Exporters	Organic Importers	Organic Processors	Organic Producers
2007	24	139	1,038	235,441
2008	342	169	1,384	405,199
2009	313	157	2,711	729,596
2010	114	153	2,454	461,774
2011	133	180	2,284	620,455
2012	800	222	2,891	685,437
2013	792	229	3,111	726,325
2014	2,266	308	5,976	901,578
2015	2,373	361	6,243	851,016
2016	2,401	367	8,637	1,108,040

Source: FiBL statistics (2018).

increased. On the other hand, the import of organic products started in 2012 and is growing slowly. The retail sale of Asian market is rapidly growing, starting with 38.7 million euros and reaching 7342.97 million euros in 2016.

Table 8.3 presents the main operators of organic outputs, those who are directly and indirectly getting benefits from expanding the market. These operators play a major role in increasing the volume of trade. There are four major stakeholders, namely, organic producers, organic exporters, organic importers and organic processors. The producer is the key stakeholder in all of them and engages in growing organic products on land. The other three enable the producer to sell his/her products in national as well as international markets. Thus, if the number of these stakeholders is increasing, it means that the market for organic products is expanding. The number of organic producers was recorded 235,441 in 2007, which reached 1,108,040 in 2016. Similarly, the number of exporters increased from 24 to 2,401; importers from 139 to 369; and processors from 1,038 to 8,637.

MAJOR ORGANIC MARKET IN ASIA

The market of organic farming is not well organized yet so there is a problem relating to information availability, especially in developing countries such as India and China in the Asian region. As per the report of the world of organic farming (2018), the Asian market for organic farming is continuously growing. It has been accounted that at least 7.3 billion euros of organic products were sold in Asia. In Asia, China reported a 5.9 billion euros market of organic products in 2016 and stands in fourth place in the world organic market. The current Indian organic market is estimated at ₹40,000 million and is likely to increase to ₹100,000–120,000 million by 2020 (Jha, 2017). In the Asian region, Japan had one of the largest domestic organic markets with a value of 1 billion euros in 2009. South Korea had a market of 281 million euros in 2015.

Table 8.4 presents the values of export and retail sale of organic products of two major players in the organic Asian market, India and China during 2007 to 2016. The export and retail values of both the

Table 8.4 *Retail and Export of Organic Product Market Value in Euros Millions*

	India		China	
Year	Export	Retail Sale	Export	Retail Sale
2007	74.46	–	–	365.41
2008	81.31	88.36	300.00	450.00
2009	87.73	92.98	300.00	790.84
2010	118.68	92.98	300.00	790.84
2011	128.41	45.90	300.00	790.84
2012	291.20	130.00	233.49	790.84
2013	291.20	130.00	364.50	2,430.00
2014	303.00	130.00	466.78	3,700.98
2015	268.58	130.00	466.78	4,712.00
2016	268.58	130.00	1,049.00	5,900.00

Source: FiBL statistics (2018).

countries are increasing, but the growth of increase of China is faster than India and the value of exports and retail have a huge gap. China began with retail value of 365.41 million euros in 2007 and reached 5,900 million euros in 2016, while India started with 88.36 million euros in 2008 and reached 130 million euros in 2016. Similarly in the case of organic exports, China started with 300 million euros in 2008 and reached 1,049 million euros in 2016, while India began with 74.46 million euros and reached 268.58 million euros in 2016.

CONCLUSION

Asia is the third largest organic market after North America and Europe, but the size of the Asian organic market is very small as compared to North America and Europe. In the global organic market, North America has captured 49.52 per cent share, followed by Europe with 39.58 per cent share and a very small share, 8.67 per cent, is captured by Asia. The other regions, Africa, Latin America and Oceania, jointly have only less than 3 per cent share of the global organic market. Within Asia, China is the fastest growing market of organic products followed by India; but the growth of China is much faster than that of India in both export and retail sale of organic products.

REFERENCES

FiBL Statistics (2018). Are data, Export and Import value data on organic agriculture worldwide world and regional 2000–2018. The Statistics. FiBL.org website maintained by the Research Institute of Organic.

Jha, D. K. (2017, 27 April). India to treble export of organic products by 2020. *Business Standard*. Retrieved from http://www.business-standard.com/article/markets/india-to-treble-export-of-organic-porducts-by–2020–117042600455_1.html (accessed on 10 March 2018).

Sahota, A. (2018). The global market for organic food and drink. In Helga Willer and Julia Lernoud (Eds.), *The world of organic agriculture* (pp. 146–149). Frick: Research Institute of Organic Agriculture and IFOAM.

United Nations. (2017). *World population prospects: The 2017 revision, key findings and advance tables* (Working Paper No. ESA/P/WP/248). New York, NY: Population Division, Department of Economic and Social Affairs, United Nations.

Willer, H., & Lernoud, J. (2018). *The world of organic agriculture. Statistics and emerging trends* (pp. 1–354). Research Institute of Organic Agriculture FiBL and IFOAM Organics International.

Organic Input and Output Market in India

INTRODUCTION

The global market for organic food was continuously increasing and reached US\$ 89.7 billion dollars (more than 80 billion euros) in 2016. In this growth, the demand for organic products was made by the developed nations. United States was the leading market with 38.9 billion euros, followed by Germany with 9.5 billion euros, France with 6.7 billion euros and China with 5.9 billion euros. In 2016, most of the major markets continued to show double-digit growth rates and the French organic market grew by 22 per cent. The highest per capita spending was recorded in Switzerland which was 274 euros per person and Denmark had the highest organic market share with 9.7 per cent of the total food market. In 2016, 2.7 million organic producers were reported. India continued to be the country with the highest number of producers, 835,200 followed by Uganda with 210,352 and Mexico with 210,000 in 2016. A total of 57.8 million hectares of land were recorded to be organically managed by the end of 2016, with a growth of 7.5 million hectares over 2015. Australia recorded the largest organic agricultural area with 27.2 million hectares, followed by Argentina with 3 million hectares and China with 2.3 million hectares (FiBL, 2018).

The growth of organic fertilizers production increased rapidly during the period 2004–2005 to 2013–2014, but states such as Assam, Gujarat, Maharashtra, Rajasthan, Tamil Nadu and Uttar Pradesh achieved slower growth rate. Assam, Kerala, Karnataka, Maharashtra, Madhya Pradesh, Tamil Nadu and West Bengal are the leading states in bio-fertilizer production and had more than 100 per cent average annual growth during 2004 to 2014. Some other states are also performing well in organic manure production. As per classification of two sub-periods in the study, the average annual growth rate of total manure production was found to be 13.07 per cent in period-I (2004–2005 to 2008–2009) and 17.67 per cent in period-II (2008–2009 to 2013–2014) in India. But the growth of major organic states varied for both the periods from –18 to 800 per cent respectively. Organic farming in Gujarat grew at 7.16 per cent rate in period-I which was lower than national average of 13.07 per cent. For period-II it recorded 33.21 per cent which was higher than national average growth of 17.67 per cent.

ORGANIC OUTPUT MARKET IN INDIA

As per the EXIM Bank's (2015) report on the organic food market, it was valued at ₹675 crores (US$150 million) in the year 2009–2010, which reached ₹1928 crores (US$306 million) in the year 2013–2014. The growth rate of organic food was recorded to be 30 per cent annually. The share of domestic and external market demand was only 31 per cent and 69 per cent respectively in 2013–2014. This trend is continuing, which means that two-thirds of organic product is exported to the rest of the world. The demand of organic products is growing because both domestic and foreign consumers are giving preference to consume organic products due to their health and test preference (Singh, 2000). The organic consumers are willing to pay a higher price for organic products such as grains, vegetables, fruits, fruit and milks. India is growing 1.24 million tons of organic food items including both edible and non-edible items. Table 9.1 presents crop-wise production of organic products in India since 2007–2008 to 2011–2012. During the Eleventh Five-year Plan period, the production increased but declined in the year 2011–2012. On the other hand, coefficient variation (CV) value of each crop is less than 2 except for dry food. This means that products whose CV value is more have more variations in production across the years and vice versa.

Table 9.2 shows the state-wise information about the organic food grain production during 2009–2010 to 2011–2012. In period-I

Table 9.1 *Category-Wise Productions of Organic Products in India (2007–2008 to 2011–2012, MT)*

Product	2007–2008	2011–2012	Period–I	SD	CV
Cotton	142,714	111,383	403,103.7	301,949.4	0.75
Cereals and millets (excluding rice)	60,044	40,786	162,200.9	136,751.2	0.84
Rice (basmati and non-basmati)	44,132	22,674	58,227.29	46,300.42	0.80
Pulses	45,518	12,957	37,753.72	15,418.2	0.41
Fruits and vegetables	10,670	8,228	395,436.8	406,087.6	1.03
Tea and coffee	11,070	6,650	63,998.4	96,161.37	1.50
Oil seeds including soyabean	68,385	2,850	193,994.2	154,629.1	0.80
Dry fruits	–	522	10,962.82	29,537.8	2.69
Medicinal and herbal plants	102,772	189	454,495.8	751,715	1.65
Miscellaneous	5,476	27	93,188.39	98,658.81	1.06
Total	490781	206266	1299070	1,623,225	1.25

Source: National Centre of Organic Farming, Various Annual Reports.

Table 9.2 *State-Wise Status of Organic Food Production (2009–2010 to 2011–2012, MT)*

State	2009–2010	2011–2012	Period–I	SD	CV
Andhra Pradesh	11,129.24	3,658.43	24,752.81	30,297.78	1.22
Arunachal Pradesh	710.02	0.00	945.77	1,083.062	1.15
Assam	2,328.89	1,200.2	6,082.013	7,499.339	1.23
Bihar	410.27	0.00	5,187.873	8,632.794	1.66
Chhattisgarh	1,278.76	3,153.66	2,042.747	984.4192	0.48
Delhi	4,765.6	0.01	2,312.623	2,385.894	1.03

(Table 9.2 Continued)

(Table 9.2 Continued)

State	2009–2010	2011–2012	Period–I	SD	CV
Goa	2,765.91	156.65	10,395.02	15,528.59	1.49
Gujarat	26,386.8	9,859.58	75,971.41	100,536.2	1.32
Haryana	3,275.85	1,731.57	41,598.94	67,719.32	1.63
Himachal Pradesh	237,105.1	472.43	104,183.6	120,990.5	1.16
J&K	12,232.56	3,513.68	8,709.73	4,593.96	0.53
Karnataka	45,472	10,324.01	92,232.44	112,807.8	1.22
Kerala	5,752.93	12,277.72	25,402.65	28,570.55	1.12
Maharashtra	53,496.16	211,740.8	319,837.4	333,786	1.04
Madhya Pradesh	164,694.5	83,404.75	489,636.3	634,517.8	1.30
Manipur	4,068.39	3.11	7,770.25	10,138.3	1.30
Meghalaya	843.56	9,654.38	8,724.193	7,459.167	0.85
Mizoram	14,473.28	0.00	63,994.1	98,572.8	1.54
Nagaland	11,120.41	560.00	6,102.627	5,299.732	0.87
Orissa	62,391.68	29,016,450	9,748,342	16,686,752	1.71
Punjab	1,970.04	0.00	23,382.62	38,806.29	1.66
Rajasthan	23,612.61	138,635.8	142,529.8	120,911.2	0.85
Sikkim	2,766.73	4,121.78	4,020.983	1,207.016	0.30
Tamil Nadu	23,847.43	19,797.66	28,428.61	11,619.82	0.41
Tripura	105.18	0	210.81	279.0455	1.32
Uttar Pradesh	970,832.7	27,526.75	430,838.5	486,279.8	1.13
Uttarakhand	10,030.05	22,439.79	37,411.63	37,200.26	0.99
West Bengal	5,561.54	3,159.97	12,371.66	13,927.16	1.13
Others	37.44	0.00	1,416.37	2,420.873	1.71
India	1,703,466	29,583,843	11,724,835	15,504,847	1.32

Source: National Centre of Organic Farming, Various Annual Reports.

presented three-year average data of organic production in India and its states. The CV value shows the variability of production across the states in the given period.

Overall, the production of all types of crops continuously increased with the increase of area but how much the Indian consumers received out of these gains is also a researchable issue. Similarly, it also needs to be investigated as to how these gains are helping in poverty elimination.

ORGANIC INPUTS MARKET IN INDIA

The growth and development of organic agriculture depends on the demand of organic products, which is continuously increasing. The supply of organic output depends on the availability of organic inputs and certified organic land. The demand for organic products is derived demand of organic inputs, growers and accreditation facility. The demand of inputs includes organic seeds, manure, vermi-compost, bio-fertilizers and accreditation mechanism. The difference between organic and conventional farming is in inputs used. In organic farming only natural seeds and fertilizer are used which are prepared naturally. While in conventional farming fertilizers and seeds are made by factories using chemicals to increase productivity. It is based on the principle live together means provide atmosphere to other living creature including flora and fauna. Soil management is done through crop rotation and usage of organic compost. Compost is prepared from domestic waste such as straw, cow and other livestock's waste and soil. Traditional method of pest control is also used to protect crops from insects, such as neem quoted seeds, kernel spray and use of ash. The entreprenerds have enormous scope for investing in organic inputs and output market management.

Nitrogen fixing and recycling via rotating crops is another method of enhancing soil fertility. This practice of enhancing soil fertility is based on the principle of low use of external fertilizer inputs. In this process of mineralization, plants release essential nitrogen to soil. Soil fertility and nitrogen level can be managed by using catch crop and green manure legumes through crop rotation. (Lockhart & Wiseman, 2012; Thoruo-Kristensen et al., 2003). The distribution of organic manure and compost via slurry injection into soil improves the power of manures. The indigenous knowledge of farmers on crops rotation and farm management can improve soil fertility. The preparation of nitrogen on farmland is costlier than industrial nitrogen. Thus, the best way to preserve nitrogen is through the recycling technique (Stolze et al., 2000).

The on farm and off-farm preparation of manure is continue in practices by the farmers and they are avoiding use of chemical fertilizer to reduce the risk of soil health. Along with the aforementioned practices, the use of different types of domestic composts, that is, ash, crop waste and manure helps in the process of stimulation in soil through micro-organisms. This process helps to maintain the organic component of soil (Fließbach & MäÈder, 2000). The use of nitrogen in organic cultivation is found to be less than 60–70 per cent than conventional method of cultivation due to slow process of recycling of residues in organic farming. So the available nitrogen in organic land needs efficient management (Kramer et al., 2006).

The management of naturally existing ingredients in soil and water is given more importance in organic farming. Organic farming requires one's own resource, that is, livestock, farmland and space for manure preparation. In traditional farming, integrated plant and nutrient management (IPNM) via humus is followed for soil and plant nutritional requirement. There is another option, that is, if farmers are unable to prepare manure at home, they can buy it from market. But the method of using organic manure and pest management is equally important. Thus, there is an opportunity for entrepreneurs to invest in organic inputs sector and get financial assistance from the government under the available schemes. Policy support and government support are available for both individual farmers and entrepreneurs. The regulation and standards are formulated to prevent illegal and unauthorized sale of organic inputs and output. The quality control of organic inputs is managed by the Fertilizer (Control) Order (FCO), 1985 and Central Insecticide Act, 1986, for the preparation and sale of organic inputs and bio-pesticides (NCOF, 2010).

The availability and production of organic inputs are essential for the growth of organic farming in India. The National Project on Organic Farming (NPOF) was launched in 2004 to increase the production of organic inputs. The Capital Investment Subsidy Scheme was stared for opening new production units of organic inputs like manure. The aim of this Scheme was to increase the production and supply of organic inputs, that is, vermicompost, bio-fertilizers, fruit, and vegetable-based, rural and urban compost. These inputs units can be established by individual or by a group of people and get finical assistance from the government. Other goals are to prevent soil, water and ecosystem degradation by using organic inputs, and set up vermiculture hatcheries for preparation of vermicompost. Thus, to achieve higher growth of organic inputs and output production (Charyulu & Biswas, 2010).

The policy document has provisions of financial assistance to set up production units for organic inputs. These financial assistances are available under back-end and credit-link subsidy schemes through National Cooperative Development Corporation (NCDC) and National Bank for Agriculture and Rural Development (NABARD). These following schemes are available for opening new production units of organic inputs: (a) Financial assistance is available at 25 per cent of the total project cost with maximum amount not exceeding ₹40 lakhs per unit for fruit/vegetables and agro-waste-based units with capacity of 100 tons per day (TPD), (b) financial assistance is available at 25 per cent of the total project cost with maximum amount not exceeding ₹20 lakhs per unit for bio-fertilizer units with capacity of 150 tons per annum (TPA) and (c) the financial burden should be shared by three stakeholders—owner (25% share), government (25% share through) and banks (remaining 50% share)—for opening vermi-culture hatcheries input units. The rate of interest is decided by the banks within 1 year of approval for finance. (Charyulu & Biswas, 2010).

The potential capacity of organic input production is higher than India's actual production. Producing at the optimum level India can fulfil both internal and external demands of organic inputs. India has various types of land and climatic advantages that can help to produce

Table 9.3 *Source of Organic Manure Production Capacity by Different Inputs*

Source of Input	Capacity (Million Ton)	% of Total Capacity
Livestock	2.47	37.60
Crop residues	2.00	30.44
City refuse	0.68	10.35
Bio-fertilizer	0.20	3.04
Biogas slurry	0.12	1.83
Green manure	0.10	1.52
Others*	1.00	15.22
Bio-pesticide**	0.001	–
All types of manure	6.57	100

Source: Bhattacharya (2006)
** Not included in the total capacity.

organically by using naturally available nutrients in soil and water. Table 9.3 presents the potential capacity of different types of organic inputs in India. Total expected production capacity of all kinds of organic inputs has been accounted to be 6.57 million tons per year using all types of resources. Livestock-based compost has the highest share in total production with 37.6 per cent, followed by crop residuals with 30.44 per cent and the city refuge contributing the 10.35 per cent in total production. All other ingredients have the remaining share in total production of organic inputs.

Bio-fertilizer Production and Growth

In the preparation of bio-fertilizer, microorganisms are used for the degradation of bio-degradable waste. These organisms are categorized into three groups: fungi, bacteria and actinomycetes. Soluble sugar in water is used for rapid bacterial proliferation. The use of biofertilizer is also beneficial to substrates proteins and amino acids by actinomycetes. The degradable ingredients such as fat, proteins, carbohydrates are mixed with other waste for degradable process through fungi organisms. The spore farming bacteria and actinomycetes are used for decomposition of cellulose in the presence of Hemicellulose in waste (Chandra, 2005). The organic and bio-fertilizers are specified in Schedule III Part A in the FCO, 1985. Bio-fertilizers also include Azotobacter, Azospirillum, Rhizobium, Mycorrhizal and phosphate solubilizing microorganisms (PSM) and are specified in Part A of Schedule IV of the FCO Act, 1985. This schedule also includes city waste compost, de-oiled cake, vermicompost and phosphate rich organic manure (PROM) as organic fertilizers.

Table 9.4 shows the annual average growth of bio-fertilizers in major states of India. The whole period is classified into two sub-periods for the analysis of annual average growth rate (AAGR) of bio-fertilizers. Period-I comprises years from 2004–2005 to 2008–2009 and period-II comprises years from 2008–2009 to 2013–2014. It is found that period-I is inconsistent in growth and production of bio-fertilizers than period-II. The value of CV shows that there was very less variation across the states during 2000–2005 to 2013–2014, except for Uttar Pradesh.

Production and Growth of Organic Compost

The preparation of organic compost uses various natural ingredients, which are available at home or farm. On the basic the content used for preparation, compost is classified into five categories: rural organic compost, urban organic compost, vermicompost, farmyard manure (FYM) and other organic compost (NPOP, 2010). Table 9.5 presents the average production and growth rate of total organic manure in India. The date is divided into two periods for comparison—period-I (2007–2008 to 2010–2011) and period-II (2010–2011 to 2013–2014). AAGR shows high and moderate variation in the growth rate across states. It is negative in the some cases, for example Andhra Pradesh, Kerala and Madhya Pradesh. The CV value has also proved this variation and the state of Assam, Gujarat and Uttar Pradesh have more than one value, which means high variation in organic compost production during the period.

Table 9.4 AAP and AAGR of Total Bio-fertilizers in Major States, 2004–2005 to 2013–2014

States	Period-I (AAP)	AAGR (%)	Period-II (AAP)	AAGR (%)	Period (I+II)	AAGR (%)	CV
Andhra Pradesh	2,691.60	3.81	1,156.81	150.57	2,023.07	85.34	0.71
Assam	68.43	254.43	114.45	10.24	89.95	118.77	0.51
Gujarat	1,195.65	7.16	2,117.96	54.04	1,753.63	33.21	0.93
Karnataka	5,387.13	338.85	7,634.32	12.56	6,082.05	157.58	0.71
Kerala	510.24	799.59	1,968.15	61.41	1,317.31	389.53	0.92
Maharashtra	2,366.26	-18.22	4,374.73	53.39	3,682.98	21.57	0.64
Madhya Pradesh	1,242.20	1.15	1,871.43	36.57	1,659.11	20.83	0.39
Orissa	223.22	116.96	521.43	38.79	383.97	73.53	0.76
Rajasthan	291.39	322.55	745.96	95.87	557.91	196.62	0.72
Tamil Nadu	2,765.20	35.40	7,686.21	63.16	5,525.54	50.83	0.79
Uttar Pradesh	392.99	122.17	2,603.16	131.79	1,669.84	127.52	1.54
West Bengal	567.79	169.11	704.99	48.70	682.77	102.21	0.81
Others	1,135.12	444.22	5,850.79	47.18	3,894.22	223.64	1.06
India	18,841.22	13.07	37,350.39	21.38	29,324.34	17.67	0.49

Source: National Centre of Organic Farming, Annual Reports.

Table 9.5 *AAP and AAGR of Total Organic Compost in Major States, 2007–2008 to 2013–2014*

States	Period–I (AAP)	AAGR (%)	Period–II (AAP)	AAGR (%)	Period (I+II)	AAGR (%)	CV
Andhra Pradesh	236.79	–12.95	98.53	–4.50	174.69	–13.91	0.76
Assam	37.91	–42.55	452.28	7,669.85	279.27	5106.30	1.52
Gujarat	32.39	979.5	284.00	224.97	175.08	624.65	1.02
Karnataka	1,679.48	107.85	1,030.75	–27.39	1,342.69	40.32	0.73
Kerala	128.92	–0.21	58.17	–47.32	88.07	–31.65	0.65
Maharashtra	69.97	2,031.34	50.43	2,839.63	55.16	2,907.9	0.87
Madhya Pradesh	81.78	14.55	103.42	–14.09	86.39	–8.70	0.69
Orissa	80.86	114.24	46.71	13.72	54.06	57.22	0.86
Rajasthan	115.66	1,879.31	242.47	1,409.6	162.57	927.88	0.96
Tamil Nadu	32.21	143.59	33.89	233.95	29.71	140.81	0.70
Uttar Pradesh	117.59	649.82	163.97	200.33	114.06	334.22	1.31
West Bengal	102.59	30.03	159.98	17.40	126.78	13.85	0.33
Others	564.84	132.5	726.73	3.62	633.99	67.87	0.38
India	3,281.04	76.5	3,451.32	30.1	3,322.57	23.23	0.24

Source: National Centre of Organic Farming, Annual Reports.

Rural Organic Compost

Rural organic compost is prepared from plant leaves, crop waste, urine and dung of cow and other livestock along with water and soil in villages. For the decomposing of the aforementioned residuals, various microbiological processes are followed, which require nearly two to three months (Misra et al., 2003). Various methods of preparation are used for decompaction of domestic biodegradable waste. Table 9.6 shows the AAGR and AAP of rural organic manure for the period-I and period-II, along with the consolidated period as a whole. It is also produced by certified agencies in six regional centres of organic farming, which show a minor decline from 945.21 million tons to 855.12 million tons in India. It happened because in some states production growth rate is negative and the CV value (between 0.04 and 1.89) also shows the variability in annual production of organic crops.

Urban Organic Compost

Urban organic compost is prepared by using household waste collected by municipal corporation from cities and towns. These wastes are composted using mechanical and non-mechanical methods. The non-mechanical method of composting is and takes several months. It is slow process of composting through bacteria and microorganisms. On the other hand, the mechanical method takes very less time to convert bio-degradable waste into compost. In the mechanical method of preparation temperature and pathogenic organisms are used for decomposing, without destroying the pathogenic organisms (Chandra, 2005). Rao (1992) has estimated potential expected capacity of urban organic compost to be 145.7 million tons per year.

Table 9.7 shows AAP and AAGR of two classified periods along with combined estimates of both the periods. The average production fell from 284.68MT in period-I (2007–2008 to 2010–2011) to 114.62MT in period-II (2010–2011 to 2013–2014). AAGR of whole period was 16.66 per cent because the growth rate was higher in period-I and negative in period-II. AAGR of Gujarat state is quite well in period-I, 33.33, per cent compared to the national average of 45.2 per cent. But

Table 9.6 *AAP and AAGR of Rural Compost in Major States, 2007–2008 to 2013–2014*

States	Period–I (AAP)	AAGR (%)	Period–II (AAP)	AAGR (%)	Period (I+II)	AAGR (%)	CV
Andhra Pradesh	48.95	1.00	48.38	0.78	48.76	1.02	0.04
Assam	2.66	288.13	0.88	–2.68	1.89	142.28	1.43
Bihar	10.45	195.85	10.12	94.32	9.39	85.8	0.89
Gujarat	5.43	–19.08	1.63	–39.31	3.10	–26.21	1.89
Karnataka	445.55	254.32	304.22	6.84	390.95	138.84	0.93
Kerala	4.10	183.33	1.55	–4,228	2.49	63.48	1.03
Maharashtra	11.35	1,186.18	7.50	–50.00	8.63	559.76	0.92
Madhya Pradesh	17.00	94.70	35.13	46.81	23.36	32.72	0.88
Orissa	46.94	67.67	35.07	32.50	34.52	36.18	0.73
Rajasthan	26.21	–59.18	16.36	–17.28	21.50	–41.11	1.09
Tamil Nadu	8.40	–24.03	3.77	–24.70	6.24	–28.58	0.93
Uttar Pradesh	7.99	12.54	3.97	–25.00	5.70	–10.40	0.71
West Bengal	41.77	65.75	78.88	42.38	57.70	33.02	0.48
Others	272.76	–17.16	207.14	3.31	243.27	–9.79	0.63
India	945.42	–2.25	750.43	–12.70	855.12	–9.44	0.54

Source: National Centre of Organic Farming, Annual Reports.

Table 9.7 *AAP and AAGR of Urban Compost in Major States, 2007–2008 to 2013–2014*

States	Period–I (AAP)	AAGR (%)	Period–II (AAP)	AAGR (%)	Period (I+II)	AAGR (%)	CV
Andhra Pradesh	1.40	41.60	5.97	4,823.03	3.88	3,215.52	1.99
Assam	0.00	0.00	0.15	2,619.69	0.08	1,746.46	2.44
Gujarat	3.19	33.33	3.13	–25.00	1.82	–33.33	2.59
Karnataka	259.66	42.95	63.84	–16.2	178.32	26.09	1.14
Kerala	3.04	3,333.33	1.78	30.36	2.18	1,686.9	0.84
Maharashtra	1.41	276.67	3.70	187.07	2.26	124.71	1.12
Madhya Pradesh	0.88	0.00	2.64	–24.64	1.51	–16.43	1.24
Orissa	0.02	6.39	0.04	87.05	0.03	59.14	0.88
Rajasthan	4.03	–33.33	12.15	–19.34	7.09	–29.56	1.21
Tamil Nadu	2.07	–29.3	2.70	383.62	2.64	239.81	1.38
Uttar Pradesh	1.88	0.00	3.75	–25.00	2.14	–16.67	1.71
West Bengal	4.03	240.83	7.53	206.24	5.56	122.49	0.64
Others	3.09	253.11	7.27	70.94	4.78	133.97	0.71
India	284.68	42.50	114.62	–27.39	212.28	16.66	0.89

Source: National Centre of Organic Farming, Annual Reports.

in period-II, it was negative. The value of CV is very high in most of the states, which means higher variability in production across the years.

Farmyard Manure

FYM is prepared using farm waste, domestic waste and livestock wastes. The ingredients such as crop bedding, straw, cow and other animal dung and urine are used for preparation of FYM. Traditional and neutral methods of preparation are used for making FYM and urine and water are used to decompose waste. An amount of 50 per cent of animal dung and remaining amount other contents is used. FYM has nearly 5–6 kg of nitrogen and potassium each and 1–2 kg phosphorus per ton. FYM quality depends on the ingredients type and animals age, along with storage and method of collection. The pit method is very popular for preparation of FYM in areas with less than 1,000mm precipitation as it easy to preserve cattle shed wastes. Another method is heap method in which cattle shed wastes are spread in uniform layers in cone shape with maximum height of 1 meter from the ground (Chandra, 2005).

Table 9.8 presents the detailed information of average production of FYM and AAGR of two periods along with the whole period, 2007–2008 to 2013–2014. Some states show positive AAGR and others are negative. Gujarat has negative growth in period-I (2007–2008 to 2010–2011), whereas it is positive in period-II (2010–2011 to 2013–2014). The overall period growth was positive. It means that in period-II the performance was better in the production of FYM. The value of CV is 1.12, which is more than 1; it means that there was huge variation in production since 2007–2008 to 2012–2013 and no consistency in production, where as in the county as whole the value of CV was 0.32, which is below 0.5 and shows less variability across the years.

Vermicompost

Vermicompost is organic manure and prepared via culturing of earthworms. Its preparation requires various ingredients such as liquid and

Table 9.8 AAP and AAGR of FYM in Major States, 2007–2008 to 2013–2014

States	Period-I (AAP)	AAGR (%)	Period-II (AAP)	AAGR (%)	Period (I+II)	AAGR (%)	CV
Andhra Pradesh	172.50	−26.97	18.80	−33.90	105.03	−36.09	1.37
Assam	26.39	−54.18	448.80	−23.84	271.15	−42.98	1.58
Gujarat	17.25	−24.44	272.25	721.04	163.71	668.47	1.12
Karnataka	861.87	352.8	537.34	−30.09	669.51	154.21	0.84
Kerala	74.49	−17.32	35.84	−36.93	54.66	33.28	0.75
Maharashtra	56.25	0.00	37.50	−50.00	42.86	33.33	0.94
Madhya Pradesh	63.40	1.17	64.13	−25.00	60.66	−16.08	0.68
Orissa	72.25	−25.92	11.13	−25.00	41.29	−29.63	1.77
Rajasthan	81.89	−33.33	207.80	−17.72	129.46	−28.48	1.10
Tamil Nadu	18.26	865.84	20.60	1,339.42	16.56	994.68	1.04
Uttar Pradesh	90.26	41,600.09	137.50	31,200	90.86	20,783	1.43
West Bengal	38.05	−0.43	36.58	5.89	37.43	−0.08	0.13
Others	177.44	381.24	257.97	133.88	189.90	203.41	0.85
India	1,750.28	12.18	2,086.22	17.15	1,873.08	3.68	0.32

Source: National Centre of Organic Farming, Annual Reports.

solid organic wastes that come from cities, pulp, paper mills, dairies, sugar industry, tanneries, food processing waste and fermentation industries. Earthworms are used as biodegraders during preparation of vermicompost. The process of vermicompost preparation is called as vermicasting, which consists of cocoons and excreta of earthworm released during the process of decomposition (Chandra, 2005).

Table 9.9 presents the AAGR and AAP of vermicompost of two classified periods, along with the whole period which gives more precise picture of growth of vermicompost across the states. All states show positive AAGR except Assam, Jharkhand, Maharashtra and Rajasthan in period-I (2007–2008 to 2010–2011) and Gujarat, Jharkhand, Karnataka and Orissa in period-II (2010–2011 to 2013–2014). The CV value of the country as a whole is 1.18, which is more than 1, which means that there is a huge variation in vermicompost production across the years. Gujarat has a value of 0.25 which shows less variability across the years. The national verticality in vermicompost production occurs due to variation in production of states such as Jharkhand, Kerala Maharashtra and Rajasthan.

Other Manure

The other manure covers animal waste and green manure. The former is prepared by using livestock excreta along with other waste and the latter is prepared by using unnecessary plants that are voluntary grown in farm and green crop straw. In rice cultivation, green manure is frequently prepared by using unnecessary plants through ploughing. In the process of ploughing these plants are mixed with water and soil, then after few days it become green manure.

Table 9.10 shows the AAGR and average production of other types of manure in major states of the country. CV value is also calculated for knowing the variation in production during the chosen periods. There was no significant change in the average production of other manure in both the periods, while in some states it declined in period-II. Thus, there was no growth in other types of manure in most of the states in India. The CV value is less than one except for Rajasthan, Karnataka and West Bengal. This means that there is no variation in production

Table 9.9 AAP and AAGR of Vermicompost in Major States, 2007–2008 to 2013–2014

States	Period–I (AAP)	AAGR (%)	Period–II (AAP)	AAGR (%)	Period (I+II)	AAGR (%)	CV
Andhra Pradesh	1.03	0.30	1.06	1.33	1.05	1.04	0.07
Assam	0.29	–33.33	1.26	55.43	0.82	20.29	0.84
Gujarat	0.73	20.00	0.58	–8.88	0.63	4.08	0.25
Jharkhand	0.10	–33.33	183.74	–25.00	105.05	–33.33	1.91
Karnataka	6.96	12.12	4.63	–10.93	5.81	–17.95	0.63
Kerala	38.54	299.68	14.40	197.6	22.39	281.57	1.18
Maharashtra	0.40	–28.04	0.49	189.71	0.48	112.45	1.13
Orissa	0.34	54.76	0.28	–31.67	0.29	6.27	0.57
Rajasthan	0.63	–33.33	0.35	873.61	0.55	565.74	1.77
Tamil Nadu	0.61	42.42	1.70	80.38	1.16	55.73	0.84
Uttar Pradesh	1.38	42.54	1.22	122.31	1.15	90.05	0.78
West Bengal	5.89	1,210.12	21.81	934.28	12.77	606.19	0.89
Others	11.91	85.71	23.16	44.36	16.35	41.55	0.53
India	68.66	2,773.4	254.66	12,713.75	168.45	9,188.5	1.18

Source: National Centre of Organic Farming, Annual Reports

Table 9.10 *AAP and AAGR of Other Manure Major States, 2007–2008 to 2013–2014*

States	Period–I (AAP)	AAGR (%)	Period–II (AAP)	AAGR (%)	Period (I+II)	AAGR (%)	CV
Andhra Pradesh	0.51	0.30	0.13	−25.00	0.29	−16.52	0.94
Gujarat	2.95	32.85	3.58	20,779.86	3.73	13,866.81	0.55
Karnataka	51.05	250.74	35.23	163.08	33.43	91.19	1.19
Kerala	8.76	4.07	3.27	−63.43	5.59	−40.25	0.75
Orissa	0.43	32.00	0.13	−47.96	0.25	−15.97	0.95
Rajasthan	0.06	−33.33	0.67	−24.81	0.42	−33.21	2.38
West Bengal	1.04	−33.33	2.25	−25.00	1.45	−33.33	1.04
Others	10.14	73.63	10.93	29.43	10.93	63.88	0.67
India	74.93	1,324.67	51.48	−778.00	53.40	−1,376.33	0.80

Source: National Centre of Organic Farming, Annual Reports.

across the years. The average production of other manure has declined from 74.93mt to 51.58mt in the chosen periods for India. The value of CV for the country as whole is 0.80, which shows some degree of variability in production.

ORGANIC OUTPUT EXTERNAL MARKET

The external market of organic products is measured by the export of organically grown food and non-food products. According to the Agricultural and Processed Food Products Export Development Authority (APEDA), Indian farmers produced around 1.35 million tons (mt) of certified organic products in 2015–2016 which included all varieties of food products namely sugarcane. Out of this 263,687mt, which was worth 298 million dollars (₹1,900 crore) was exported.

The government imposes quantitative restrictions on export of some products for ensuring food security at home. Menon, executive director of the Indian Competence Centre for Organic Agriculture—a

Table 9.11 *Export of Organic Products from India (2014–2015 to 2016–2017)*

Year	Export Quantity (in MT)	Annual Growth Rate (%)	Export Value (in Rs. Crore)	Export Value (in USD million)
2007–2008	37,533	–	498.22	100.4
2008–2009	44,476	28.50	537.00	116.0
2009–2010	58,408	31.32	526.50	112.0
2010–2011	69,837	19.57	699.00	–
2011–2012	157,800	125.95	839.34	–
2012–2013	165,262	4.73	1,155.81	–
2013–2014	194,088	17.44	1,328.61	220
2014–2015	285,663	47.18	2,099.16	327
2015–2016	263,687	–7.69	1,975.87	298
2016–2017	309,767	17.48	2,478.17	370

Source: APEDA.

Bengaluru-based network, believes that the overall market of ₹4,000 crore under the organic value chain would hit ₹10,000 to ₹12,000 crore by 2020 with similar increase in export. While the export of organic wheat, non-basmati rice, edible oils and sugar have been exempted from all annual quantitative ceilings with immediate effect, those on pulses and lentils has been increased from 10,000 tons to 50,000 tons. Farmers export largely to Europe, Canada and West Asia. Oilseeds were half of India's overall organic export, followed by processed food products at 25 per cent.

The volume and value of organic export are presented state wise in Table 9.12. There were 6,674 certified operators under National Programme for Organic Production (NPOP) as on 31 March 2017, which included individual producers, processors, traders, grower groups and wild collection operators. Madhya Pradesh is the largest exporter of organic product and Gujarat stands in second position.

Table 9.12 *State-wise Export Organic Products under NPOP in 2016–2017*

State Name	Exported Qty (MT)	% of Total Export	Total Value in ₹ (in Lac)
Andhra Pradesh	348	0.1	11,187.39
Assam	11	0.0	51.79
Chhattisgarh	32	0.0	290.41
Goa	442	0.1	18,190.45
Gujarat	53,327	17.2	33,293.45
Haryana	5,468	1.8	3,507.07
Himachal Pradesh	9	0.0	14.61
Jammu and Kashmir	821	0.3	1,047.34
Jharkhand	0.3	0.0	6.33
Karnataka	8,538	2.8	16,526.52
Kerala	4,282	1.4	17,649.10
Madhya Pradesh	141,346	45.6	62,599.93

(Table 9.12 Continued)

(Table 9.12 Continued)

State Name	Exported Qty (MT)	% of Total Export	Total Value in ₹ (in Lac)
Maharashtra	31,629	10.2	18,967.72
Meghalaya	1	0.0	19.61
New Delhi	45,720	14.8	26,364.51
Odisha	9	0.0	18.25
Punjab	205	0.1	171.95
Rajasthan	5,760	1.9	5,344.89
Tamil Nadu	1,402	0.5	3,108.23
Telangana	3,511	1.1	4,048.14
Uttar Pradesh	1,999	0.6	3,395.81
Uttarakhand	446	0.1	645.60
West Bengal	4,461	1.4	21,368.60
Total	309,767	100	247,817.67

Source: APEDA.

Delhi and Maharashtra are other states which contribute more than 25 per cent of the total exports. All these four states contributed more than 85 per cent of total organic states.

CONCLUSION

The growth of organic inputs and output market is showing upward trend in India. The organic output market is not developed in many states in India and is not accessible for small and marginal organic producers; only large farmers are selling their organic products both in the niche and national market. On the other hand, the organic inputs market is not much popular and developed in all states. Farmers are preparing their own compost on their farm lands and some government sponsored units are selling organic inputs to producers. The overseas market for organic output is gathering momentum and some products are already very famous across the world. The volume of organic export was 309,767mt in 2016–2017.

REFERENCES

Chandra, K. (2005). *Organic manures*. Bangalore: Regional Centre of Organic Farming.

Charyulu, K. & Biswas, S. (2010). *Organic input production and marketing in India: efficiency, issues and policies* (CMA Publication No 239). IIM Ahmadabad.

EXIM Bank. (2015). *Potential for trade of organic products from India* (Occasional Paper No. 174). Mumbai: Author.

Fließbach, A., & MaÈder, P. (2000). Microbial biomass and size-density fractions differ between soils of organic and conventional agricultural systems. *Soil Biology and Biochemistry, 32*(6), 757–768.

Kramer, S. B., Reganold, J. P., Glover, J. D., Bohannan, B. J., & Mooney, H. A. (2006). Reduced nitrate leaching and enhanced denitrifier activity and efficiency in organically fertilized soils. *Proceedings of the National Academy of Sciences of the United States of America, 103*(12), 4522–4527.

Lockhart, J. A. R., & Wiseman, A. J. L. (2012). *Introduction to crop husbandry: Including grassland*. Elsevier.

Misra, R. V., Roy, R. N., & Hiraoka, H. (2003). *On-farm composting methods*. Food and agriculture organization of the United Nations (FAO).

NPOP (2005). *National program for organic farming*. Department of Commerce, Ministry of Commerce & Industries, Government of India, 6th Edition.

Singh, R. B. (2000). Environmental consequences of agricultural development: a case study from the Green Revolution state of Haryana, India. *Agriculture, Ecosystems & Environment, 82*(1), 97–103.

Stolze, M., Piorr, A., Häring, A. M., & Dabbert, S. (2000). *Environmental impacts of organic farming in Europe*. Universität Hohenheim, Stuttgart-Hohenheim.

Thorup-Kristensen, K., Magid, J., & Jensen, L. S. (2003). Catch crops and green manures as biological tools in nitrogen management in temperate zones. *Advances in Agronomy, 79*(79), 227–302.

Descriptive
Analysis

Organic Farming in Asia

INTRODUCTION

World agriculture is moving towards organic farming and a total of 58.7 million hectares of land was recorded as organic farmland in 2016, which is 1.2 per cent of the total agricultural land. The organic farmland increased by 7.5 million hectares with a growth of 15 per cent from 2015. This is mainly due to Australia adding 5 million hectares in 2016. The remaining 2.5 million hectares of land were added by the rest of the world. The regional growth varies from the above region. In Europe the area added was 1 million hectares with growth of 6.7 per cent; Asia added 0.9 million hectares with growth of 34 per cent; Africa added 0.1 million hectares with 7 per cent growth; Latin America added 0.4 million hectares with growth of 6 per cent; and North America added 0.2 million hectares with growth of 5 per cent.

Asia has been on third place after Oceania and Europe in terms of organic agricultural land (OAL) and organic output since 2000. This region has a huge opportunity in production of organic food grains, vegetables, pulses, spices, cotton, herbs and wild products. But the domestic market of organic output is not well developed, and only metropolitan cities have consumers of organic products. In this chapter, we have tried to present the status and growth of organic farming in Asia. The countries which are leading in production of organic products are China, India, Kazakhstan, Philippines and Indonesia. The growth of organic farming within the Asian region and the leading countries growing organic products is also presented in this chapter.

STATUS AND GROWTH OF ORGANIC FARMING IN ASIA

The journey of organic farming began in the 2000s when certified organic producers were recorded in most of the Asian counties, but it was practiced for achieving self-sufficiency in food, improve soil fertility and for export purpose. Data on area under organic farming in most of the counties are not available before 2000. In the 1990s, the Organic Bank in Japan promotes organic agriculture with its campaign "1% organic", which aims to cultivate at least one per cent organically of the total land. After 2000 the annual report on the world of organic farming began to publish which provided regional as well as countries-wise data on organic farming. In the first report, the area under organic farming in Asia was recorded as 60,532 hectares (Willer & Yussefi, 2000).

In 2005, the area under organic farming was still small in Asia but was increasing rapidly. The major organic producer countries in Asia were China, India, Indonesia and Japan. The data on organic farming in other Asian countries are still not available. As per the report of the world of organic agriculture (2005), the total organic area under organic farming was recorded to be 736,000 hectares, managed by 66,000 farms and 2.9 million hectares were certified as 'wild harvested' areas in Asia. In Asia due to lack of national certification and organic regulation, many countries are unable attract foreign organic consumers (Willer & Yussefi, 2005).

The total organic agricultural area under organic farming was recorded to be nearly 3.3 million hectares in Asia in 2010. The share of Asian countries in the world's OAL is 9 per cent only with 4 million producers. In Asia, China is still the leading country with an organic agricultural area of 1.9 million hectares, followed by India with 1 million hectares of OAL. Another country in Asia is Timor Leste which has the highest organic agricultural area as a proportion of total agricultural land (TAL) 7 per cent. The organic agricultural area also includes organic wild collection areas which play a major role in India and China. Organic aquaculture is also important in growth of organic farming in China, Bangladesh and Thailand. The sale of organic output in Asia is still depends on the export, which support the growth of organic farming in the region. (Willer & Yussefi, 2010).

As per FiBL survey 2016 and Willer et al. (2018), Asia has recorded 4.9 million hectares of OAL in 2016. This constitutes 9 per cent of the world's OAL. China is the leading country with the highest area of OAL with 2.3 million hectares; India stands second with an area of 1.49 million hectares. Timor-Leste has the highest proportion of OAL (almost 7%). In Asia 19 counties have regulations on organic agriculture and another 5 are in the process of drafting regulations. The sale of organic food produced in Asian countries is continuously increasing and China is still dominant in terms of market and organic production. India aside from being an exporter, has a growing demand of organic foods from urban domestic market.

Table 10.1 shows the increase of certified organic agricultural area since 2000 to 2016. This area covered both cultivated organic land and wild collection and beekeeping area, from where the herbs and other forest products are collected. In 2000, both Africa and Europe region had negligible area under organic farming which constitutes nearly zero percent and one per cent of total world organic cultivated land. In the same year Europe, Latin America and North America showed their presence in organic farming and had less than 1 per cent land of TAL under organic farming, while Oceania region recorded more than 1 per cent land of TAL under organic farming. After a decade, in 2010, Africa, Asia and North America had less than 1 per cent land of total agriculture land under organic farming, But the other regions, Europe and Oceania, expanded their area under organic farming and reached more than 2 per cent of TAL, while Latin America excide than 1 per cent of total agriculture land.

Most recent data for the year 2016 showed that only Latin America's share declined from 1.11 per cent to 0.94 per cent during 2010 to 2016. It happened because farmers quit organic farming and shifted to conventional farming due to fall in income and output. Other three regions, Africa, Asia and North America were still below 1 per cent area of the total agriculture land under organic farming. Oceania attained the highest growth and recorded 6.51 per cent area under organic farming to total agricultural land, followed by Europe with 2.70 per cent.

Table 10.2 presents growth of organic agriculture in the Asian region from 2000 to 2016. This table shows the number of organic

Table 10.1 *Regional Increase of Organic Agriculture Area (ha)*

Region/year	2000	% of TAL	2010	% of TAL	2016	% of TAL
Africa	52,675.50	0.0	1,075,832.55	0.09	1,801,699.06	0.16
Asia	60,532.11	0.0	2,457,914.99	0.15	4,897,837.45	0.30
Europe	4,581,068.22	0.89	10,028,781.07	2.07	13,534,411.24	2.70
Latin America	3,910,608.00	0.56	7,539,643.36	1.11	7,135,135.30	0.94
North America	1,058,951.00	0.27	2,472,679.00	0.63	3,130,331.86	0.77
Oceania	5,310,157.00	1.10	12,145,055.40	2.87	27,346,985.81	6.51
World	14,973,991.83	–	35,717,635.38	–	57,842,023.74	–

Source: FiBL statistics (2018).

Table 10.2 *Growth of Organic Agriculture in Asia*

Year	Number of Producers	Growth of Organic Producers	OAL Area (ha)	Growth of OAL	% OAL to TAL
2000	5,289	–	60,532.11	–	–
2001	59,568	1,026.26	420,199.47	594.18	0.03
2002	61,721	3.61	429,325.67	2.17	0.03
2003	13,966	–77.37	494,779.80	15.25	0.03
2004	100,177	617.29	3,781,817.79	664.34	0.24
2005	168,697	68.40	2,678,703.62	–29.17	0.17
2006	194,521	15.31	3,001,261.65	12.04	0.19
2007	235,441	21.04	2,902,697.19	–3.28	0.18
2008	405,199	72.10	3,359,183.43	15.73	0.21
2009	729,596	80.06	3,580,459.78	6.59	0.23
2010	461,774	–36.71	2,457,914.99	–31.35	0.15
2011	620,455	34.36	3,692,387.47	50.22	0.23
2012	685,437	10.47	3,218,701.15	–12.83	0.20
2013	726,325	5.97	3,408,911.67	5.91	0.21
2014	901,578	24.13	3,567,578.40	4.65	0.22
2015	851,016	–5.61	3,965,288.91	11.15	0.25
2016	1,108,040	30.20	4,897,837.45	23.52	0.30

Source: FiBL statistics (2018).

producers, OAL in hectare and per cent of TAL. Years 2002 and 2004 show three-digit growth of organic producers from the previous years, and year 2003, 2010 and 2015 show negative growth of organic producers. It means these years the number organic producers fell. In 2002 and 2013 shows one-digit growth from previous year and rest of the other years recoded double digit growth of organic producers in Asia. Similar result shows in case OFL growth. In year 2005, 2007, 2010 and 2012 registered negative growth from the previous year. It means the area under organic farming have decline from the previous year. The year 2001 and 2004 shows three-digit growth in organic farmland similar in case producers and in year 2003, 2009, 2013 and 2014 show one digit and rest of the years recorded two-digit growth.

MAJOR ORGANIC PRODUCER COUNTRIES IN ASIA

According to the annual report of FiBL and IFOAM (Willer & Lernoud, 2017), *The World of Organic Agriculture: Statistics and Emerging Trend 2017,* Asia has 35 per cent of the total organic producers in the world, followed by Africa with 30 per cent and Latin America with 19 per cen at third place. The reason Asia is at the first place is that the two highest populous countries China and India are leading in organic farming and the landholdings of farmers are very small. In Asia, out of the 39 organic farming countries only 19 countries have organic farming rules and regulations, and 5 are in the process of drafting regulations which are essential for the growth of organic farming.

There are five major leading countries in Asia namely China, India, Kazakhstan, Philippines and Indonesia. The growth of organic farming and country position is measured by the area under organic farming and number of organic producers. Table 10.3 shows the area under organic farming in five major countries in Asia. China and India are leading nations in Asia but the share of OAL to TAL is still below 1 per cent. In Asia, Philippines is the only country which had an OAL nearly 2 per cent of TAL in 2016. As per Table 10.3, China has 2.3 million hectares of organic land, followed by India with 1.5 million hectares; rest of all other countries are still below 1 million hectares of organic land in 2016.

Table 10.3 *Increase Organic Farmland in Major Asian Counties (ha)*

Year	China	India	Kazakhstan	Philippines	Indonesia
2000	40,000	2,775	–	2,000	–
2001	301,295	41,000	–	2,000	40,000
2002	301,295	37,050	–	2,000	40,000
2003	298,990	73,500	–	3,500	40,000
2004	3,466,570	114,037	–	14,134	52,882
2005	2,300,000	185,937	–	14,134	17,783
2006	2,300,000	432,259	2,393	5,691	40,419
2007	1,553,000	1,030,311	2,393	15,344	69,606
2008	1,853,000	1,018,469	157,176	15,794	54,509
2009	1,853,000	1,180,000	134,861	51,806	59,141
2010	1,090,000	780,000	133,561	79,992	71,114
2011	1,900,000	1,084,266	196,215	96,317	74,034
2012	1,900,000	500,000	291,203	80,974	88,247
2013	2,094,000	510,000	291,203	86,155	65,688
2014	1,925,000	720,000	291,203	110,084	113,638
2015	1,609,928	1,180,000	303,381	234,642	130,384
2016	2,281,215	1,490,000	303,381	198,309	126,014

Source: FiBL statistics (2018).

CONCLUSION

Asia is one of the leading regions in organic agriculture with an area of 4.9 million hectares after Europe with 13.5 million hectares and Oceania with 27.3 million hectares. Although Asia has the largest number of organic producers in the world, the area and production are still not increasing as in the other regions. The reason behind the large number of producers in Asia is that the two most populous countries, China and India, have smaller farmland proportions. Except Philippines, Bhutan, Iraq, Israel and occupied Palestinian Territory, all Asian countries have less than 1 per cent of TAL area under organic farming.

REFERENCES

FiBL Statistics (2018): Are data on organic agriculture worldwide world and regional 2000–2018. The Statistics. FiBL.org website maintained by the Research Institute of Organic.

Willer, H., & Yussefi, M. (Eds.). (2000). *Organic agriculture world-wide: statistics and perspectives.* Bad Durkheim: Stiftung Ökologie & Landbau (SÖL).

Willer, H., & Yussefi M. (Eds.). (2005). *The world of organic agriculture: Statistics and emerging trends 2005.* Bonn: Research Institute of Organic Agriculture (FiBL), Frick: IFOAM Organics International.

Willer, H., & Lukas Kilcher (Eds.). (2010). *The world of organic agriculture: Statistics and emerging trends 2010.* Bonn: Research Institute of Organic Agriculture (FiBL), Frick: IFOAM Organics International.

Willer, H., & Lernoud, J. (Eds.). (2017). *The world of organic agriculture. Statistics and emerging trends 2017.* Bonn: Research Institute of Organic Agriculture (FiBL), Frick: IFOAM Organics International.

Willer, H., Lernoud, J., Huber, B., & Sahota, A. (Eds). (2018). *The world of organic agriculture. Statistics and emerging trends 2018.* Bonn: Research Institute of Organic Agriculture (FiBL), Frick: IFOAM Organics International.

Organic Farming in India

INTRODUCTION

India attained self-sufficiency in production of food grains after the Green Revolution. This was possible due to mechanization of agriculture along with the use of chemical fertilizers and high yielding variety seeds. As per economic surveys, India produced 50.82mt of food grains in 1950–1951, 74.33mt in 1966–1967, 212.05mt in 2004–2005, 259.32mt in 2011–2012 and 281.37mt in 2018–2019. India realized that production of food grains met the demand of its people, but the unblamed use of chemical fertilizers and pesticides had adversely affected the ecosystem. These above challenges opened the door for organic farming in India and India began to participate in the worldwide organic movement in the 2000s. The organic movement was initiated by the civil society, private organizations, research institutes and NGOs.

Organic agriculture is a holistic approach of cultivation in which manure, plants, compost and bio-organisms are used. It is a method of cultivation based on the principle of health, ecology, care and fairness. It allows the use of only those inputs which enrich soil, product and nature. Its nomenclature varies and can be called biological, ecological, natural or organic farming (Baker, 2010). As per the Codex Alimentarius Commission, it is a holistic management production system which enhances the agro-ecosystem. Integrated management of the production system means to integrate ecological, social and economic issues with farming.

The mechanization of agriculture and the use of chemical fertilizers have depleted the ecosystem and natural resources: soil, groundwater and water bodies. The release of ammonia and nitrous oxide in the atmosphere is the cause of depletion of the ozone layer and acid rain (Curtzon & Enhalt, 1977; Lægreid, Bockman, & Kaarstad, 1999; Prasad, 2005). Marwaha (2005) has conducted a study on the use of pesticides in India. He found that the use of pesticides remains low at 480 g ha^{-1} as compared to 10,770 g ha^{-1} in Japan, 4,000 g ha^{-1} in the USA and 2,500 g ha^{-1} in Europe. The unbalanced use of chemical fertilizers and pesticides leads to degradation of natural resources.

For the above reason, the movement of organic farming is gaining popularity across counties. To explore this opportunity, India participated in the worldwide organic movement and tried to learn from other countries' experiences. India set up a task force in 2000 to explore the opportunity and possibility of organic farming. The task force was headed by B. K. Yadav, Director of Gujrat Agriculture Department. The task force reported on the possibility of benefits and increased the awareness of organic farming. This task force set up a steering committee headed by M. S. Swaminathan for suggesting organic farming feasibility. The committee suggested it is as a thrust area and it should be first implemented only in rainfed regions of the country. Later in this direction, APEDA and Ministry of Commerce launched NPOP in 2000.

The aim of NPOP was to provide information to the consumers on organic products and producers on the method of organic cultivation. The process of accreditation and certification are the first steps towards organic farming. The third-party certification system is the most popular in many countries and is expensive for farmers. Further, the alternative cost-effective participatory guarantee system (PGS) was developed by IFOAM based on group participation. India launched this system in 2011 with the name Participatory Guarantee System India (PGS-India). These public policy incentives increase certified area and number of farmers. In 2000 there were only 1,426 certified farms with an area of 2,775 hectares in India (Sujatha, 2008). The area under organic farming increased to 42,000 hectares in 2003–2004. In 2018, 3.56 million hectares of certified land with 835,000 producers were recorded (APEDA, 2019).

High growth in organic farming is recorded in India which possible because of the joint effort of people and the government. Some states are performing better than others due to many reasons, such as climatic conditions, policy incentives, and people's knowledge on organic farming and products. Madhya Pradesh is the leading state in organic farming having 1.1 million hectares certified area, which is 52 per cent of the total certified area. Maharashtra is at second place with 0.96 million hectares land and 33.6 per cent of total certified area followed by Orissa and other north-eastern states (Shukla et al., 2013).

Two approaches are popular for organic farming: One is profit-centric and the other is the livelihood approach. The former is adopted by commercial growers to produce cash crops such as cotton, basmati rice, dry fruits and spices, for the purpose of export and profit. The latter was adopted by resource-poor farmers who cultivate local crops to sell niche market. In India most of the farmers fall under the former category. The high demand of organic products attracts farmers to adopt organic farming to sell their product at a higher price (Yadav, 2008).

This chapter is divided into three subsections. The first subsection talks about the early history of organic farming in the world and how people across counties have taken initiative for the worldwide organic movement. Then the second section illustrates the growth of organic farming in India. The growth of certified area is the indicator for growth of organic farming. Production data of different kinds of organic products is not available. Section three discusses about the growth of clusters in organic farming. Only a few states are dominating the production in organic farming.

HISTORY OF ORGANIC FARMING

The history of organic farming is rooted with traditional farming practices. The modern conventional method of cultivation is a means to enhance production and productivity. Various historical evidences are available for the existence of organic farming, but the nomenclature was different. Here presented is a detailed review on development of organic farming worldwide.

Evidence form the World

The concept of organic farming emerged in the 1900s, but its history deep rooted in the organic movement of civil society. Its modern form began in USA United States and Europe in the early 20th century. The early effort on organic farming was motivated to resolve ecological problem which are increasing under conventional farming. These problems are; depletion of soil and water quality, quality of food, quality of livestock, human health and rural poverty. So, the organic farming is popularized as solution of theses problem and maintain the health of ecology. The sustainability of agriculture depends on the health of soil, which further depends on the humus. The management of organic component of soil is called humus faming using traditional method of cultivation (Kuepper, 2010).

An alternative method of cultivation was developed in United States and Europe during 1920s. This was developed due to degradation of environmental resources under existing method of cultivation. These degradations are; decline of ground water, depletion of soil, quality of foods and reduction in livestock. The concept of organic farming was developed in response to above mentioned problems. It is based on the concept of health as whole ecosystem (Kuepper, 2010). The research on soil quality, water and other natural resource management helps to develop this framing with scientific provisions. The use of humans in organic farming was the primary development the direction of organic farming, which was known as biodynamic farming. In 1924, Rudolf Steiner was Austrian Economist who developed the biodynamic agriculture, which was the first movement towards organic farming.

The focus of organic farming movement was shifted towards the method and process of cultivation of crops without using chemical fertilizers during 1930s and 1940s. The growth of plants should be maintained by using organic composts. This academic debate on alternative method of cultivation helps to develop an 'alternative agriculture' (Heaton, 2001).[1] The different associations are established for

---

[1] Lady Eve Balfour conducted a long-term project of comparing organic and non-organic production in the 1940s at UK and published a book, *The Living Soil*, pointing towards the importance of healthy soil and nutritional superiority of organically grown food.

the development of this method of alternative agriculture, that is, Soil Association (1946)[2] in United Kingdom, Soil and Health Association (1941)[3] in New Zealand and Rodale Institute (1946)[4] in United States of America. In this period humans farming becomes popular for soil management in western countries. This organic movement encourage Japan to do innovation on agriculture and in 1936 Mokechi Okada formulated the concept of nature farming. Nature farming refers to agronomics and spiritual aspect of cultivation and humanity.

The development of organic agriculture is deep rooted with bio-dynamic farming and followers of Rudolf Seiner. Albert Howard was a pioneer who carried organic farming in the west and associated with Indian agriculture. He has written about traditional practices in Indian agriculture and indoor method of composting in his book *An agriculture Treatment* in 1943. In this period humus farming was replaced with organic farming. The term 'organic farming' was used first time by Lord Northbourne in his book *Look to the Land* in 1940s. He defined as 'the farm must have biological completeness, it must be living entity, it must be itself a balanced organic life' (Bacon, 2005).

The divergence of organic agriculture from mainstream agriculture happen during 1960s. In this period the three events have significant impact on the growth of organic agriculture. One is the counter cultural movement in United State of America (1960–1970), which support this organic movement both politically and socially. This movement helps to increase the demand of organic products that leads to increase the production and productivity. The people have seen the regain of environment under organic farming (Haedicke, 2012). The book *Silent Spring* was written by Rachel Carson stated that scientific certainty and impact of pesticides on environmental domains. It also emphasizes the damage of ecology and provoke against the

---

[2] The Soil Association was founded in 1946 by a group of individuals who were concerned about the health implications of increasingly intensive agricultural systems.

[3] The *Soil* and *Health Association* was *founded* by Dr Guy Chapman in 1941. It promotes organic food and farming in *New Zealand*.

[4] Rodale Institute was founded by organic pioneer J. I. Rodale in 1947. The aim of institute was to study the link between healthy soil, healthy food and healthy people.

conventional farming. This book combines the both organic and environmental movement together (Beyl, 1992).

The use of Dichloro Diphenyl Trichloroethane (DDT) has banned in developed countries during 1970s but the but its export to developing countries continue and farmers use in agriculture practices. Rachel Carson (1962) has warn the farmers about the dangers of pesticides for ecosystem. The movement of organic farming was supported by *Back to Landers Movement*[5] in this period and younger generation migrate from urban to rural area (Jacob, 1997). The formation of IFOAM at Versailles in France in 1972 was a landmark achievement in promotion to organic movement. It helps to unite people, society and political parties to support organic movement (Neggli & Lockeretz, 1996). It was the biggest international NGOs working for promotion of organic farming worldwide. The efforts of this institution and other institutions, which are supporting organic movement and spreading the information about the organic farming across the countries. It facilitates accreditation and certification certificate for all stakeholders of organic farming.

After the establishment of IFOAM, the organic farming emerges as subsector of agriculture during 1980s. The western countries came together and developed method of certification, that is, California Certified Organic Farmers (CCOF) in 1973. It was the earliest accreditation and cortication organization in the world (Guthman, 2002). The institutionalization of organic farming begins by passing legislative rule and regulation. These legislations were response of government to people of organic supporters. Both consumers and producers force their governments make rules and regulations for organic agriculture in Demark, France, Spain and Europe (Graf & Willer, 2001). In this period movement also begins to knock the door of developing countries like, Argentina, Cuba, Mexico and South America. All these countries begin to formulate their own accreditation rules and regulation for organic farming and products. The domestic demand of quality food

---

[5] The 'Back to Landers' has emerged as vigourous concept in 1960–1970s in USA; the books such as *At Home in the Woods: Living the Life of Thoreau Today* (1951), *Living the Good Life* (1954) and *We Like it Wild* (1963) narrated the story of self-sufficiency and simple life and they became influential for the growth and expansion of Back to Lander Movement.

in Cuba force to search alternative method of cultivation. The debate on organic versus conventional farming has been observed across the world in this period. The lesser research was conducted on organic farming method, principle and products (Lockeretz, 2002).

The new agrarian discourse begins during 1990s and started to blame that organic farming following the path of conventional farming (Kristiansen et al., 2006). The market of organic products was extended and replacing the local veritable venders. In this period the demand was increasing for organic products like food grain, fruits, vegetables, milks for both internal and external market. The accreditation and certification standards are framed and implemented by the producers, processor and traders. The regulation for the trade of organic products are also formalized. The international acceptance of organic farming was recognized in this period and defined differently. The basic principle and main idea remain unchanged (IFOAM, 2002).

The producers, processors, traders and NGOs raised the issues of cost of certification under third parity certification system during 2000s. The third-party certification system was operated by private certification agencies after getting permission from IFOAM and other organic national bodies. The IFOAM recognized these three certification systems, (a) Organic Guarantee System of IFOAM (b) Third-Party Certification System and (c) Participatory Guarantee System (PGS). All these methods of certification were accepted for international trade of organic products. The ideology, perspective and movement on organic farming has been replaced with certification mechanism, standards, technology for packaging and market networking (Khosla, 2006). In this period the production and growth of certified organic area was increased across the counties. But the growth of Asian region was recorded not much as other regions of the world because they are constrained with food security issue (Singh, 2010).

Evidence from India

The evidence of farming in India was found science ancient time period when people started to live in the group. The agriculture mainly based on traditional method of cultivation. The several evidences found that people use cow dung, waste of plant leaf and other home waste manure. In the literature of accident India mentioned the use of domestic manure. The description on manure was found in Rig Veda and Sukra in verses 1, 10 and 161 during 2500–1500 BC. After that Atharva Veda II in Sukra (4,5, 94, 107–112). These evidences found that the concept agriculture based on farming with nature. In the literature found that science of plants is called 'Vrikshayuved', science of agriculture is called; Krishisastra', and science of animal called 'Mrugayurveda' (Mahale and Soree, 1999; Pereira, 1993; Randhawa, 1986). In the Arthashtra the book of Kuutilya (300 BC) also discuss about the excreta of aminol and oil cake. The Brihad Sansitha written by Varahmihir described the method of manure preparation form crop straw. From above these facts the origin of agriculture begins with nature (Behera et al., 2012).

The green revolution and liberalization police made distraction of natural resources through using chemical fertilizers and pesticides in farming (Grigg, 1989). Howard pointed out that India was discoverer of nature friendly farming and practicing from centuries (Narayanan, 2005). In India, agriculture farming shifted from tradition to conventional method of cultivation during 1970s. The use of chemical fertilizers, HYV seeds and pesticides increases the production of wheat and others crops and able to meet the growing domestic demand of food. But along with this positive effect of conventional farming the negative impact on natural resource, increasing the cost of production and the issue of quality product are also occurred. Increasing the cost of cultivation was the cause of financial distress among the peasant Indian farmers.

The academic effort on organic farming begins during 1980s, with the first conference on organic farming held at Wardha, Maharashtra in 1984 organized by Association for propagation of Indigenous Resources (APIGR). Then another conference held at Bordi, Maharashtra on organic farming and paste management in 1989. Further, the government of Rajasthan and Rajasthan College of Agriculture held a meeting on organic agriculture in 1992. In the line in 1993 and 1995 organized two national conferences on organic farming. The outcome of conferences was to establishment of Agricultural

Renewal Institute for Sustainable Environment (AIRSE) in 1995. The aim of this institute to promote organic farming through support of regional groups and research in the country (Narayanan, 2005).

There were three types of farmers practicing organic cultivation during 1990s in India. One those who are educated and based urban areas influenced with this method of cultivation technique. They did not to wish stay with organic farming in the long run. The second group is educated and having scientific knowledge on organic farming method. The third is group of farmers are practice organic farming based on trial and error method. But it has been found that only those farmers sustain in organic farming, having their own land, water and livestock. They are producing cash crops, that is, coconut, pepper, sugarcane, cotton, spices, etc. many other quite organic farming those income and yields falls (Narayanan, 2005). Thus, the opting organic farming method is directly depending on the profit of the farmers and their knowledge of it.

The period of 2000s was crucial for growth and development of organic farming in India. The movement of organic farming in true spirit riches in this in this period when, the government formulate rules and regulation on organic farming. The first indication this direction was the launching the scheme of National Program for Organic Production in May 2000. This scheme was a land make in the history of organic farming in India. The different third-party certification agencies start to accredited and certified organic land and products. The NPOP also launched 'India Organic' logo for organic products in 2002. Then the NPOP 2005 policy document on organic farming make the provision for accreditation, certification and other provisions for organic farming.

SCHOOLS OF THOUGHT ON ORGANIC FARMING

The early thinker on organic farming was not philosopher but he was a botanist Albert Howard (1873–1947) in England. Who travelled India and Asia and documented in his book *Agriculture Testament* (1940) observing different method of agriculture practiced that time by using manure and composting in organic farming? He establishes foundation for organic movement and argued that soil fertility is the foundation of sustainable agriculture. After that Eve Balfour (1899–1990) was a British lady farmer who recognized lack of scientific method for organic farming during her organic agriculture research project titled as Haughley Experiment. At the same time Austrian philosopher Rudolph Steiner (1861–1925) presented their claim on organic farming based on observation, tradition and anecdotal evidence.

The scientific and experimental research on organic techniques and farming method begin in 1940's after the establishment of Rodale Institute founded by J.I. Rodale (1898–1971) in Pennsylvania. The Institute identified best practices in organic farming and applied scientific for scientific proof. Rodale Institute is leading institute in organic research and later in 1990's research begins in reputed universities of the world. But the Research on organic farming in academics begins 2000's in India, when the organic movement spread across the world and reaches in India. The historical and traditional knowledge on organic farming was deep rooted in Indian agriculture system. There were many reasons for the that; Indian farming was labour and livestock intensive, tradition seed are used, land holding are small, animal are used for cultivation and manure and compost are use as nutrients supplements. In India we can calcify organic farming into three schools of thought: traditional thought, natural thought and scientific thought.

Indigenous farmers propound the traditional school of thought, which is based on the indigenous culture, tradition, techniques and knowledge. Since the ancient time Indian are practicing this type farming and cultivated many crops across the India without any scientific and product certification. This type of forming is popularly known as *Indigenous Agriculture* based on non-western techniques and knowledge. Another Natural School of Thought is based on both traditional and scientific knowledge, but the traditional and farmers thoughts given more weight, the philosophers and environmentalist's and social scientists are the supporter of this thoughts. In this school of thought, Pfeiffer called *biodynamic agriculture*, Hans Mueler Switzerland politician in 1930, imputed the name as *organic-biological agriculture*, Mokicho Okada a Japanese farmer coined as *Nature Farming* based on the power of water, earth and fire in mid 1940s, Bill

Millision coined the name of *Permaculture* in late 1970s and advocated perennial crops. Natural farming and Permaculture widely practiced in India. The Scientific School of Thought provide to scientific validity of organic farming and the phrase *sustainable agriculture* was coined by Australian agricultural scientist Gordon McClymont and W. Jackson as cited in the book *New Roots for Agriculture* in 1980.

STATUS AND GROWTH OF ORGANIC FARMING

The growth of organic farming is measured via the certified organic area of the cultivated land. The production of organic inputs and output measure strength of market. The certified organic area further classified into two categories, one is fully converted certified area, and another is certified area under the conversion period. In India one additional category is area under the wildlife collection of organic products. The forest cover area from where forest and wildlife products are collected comes under this category. But these areas also must follow the organic standards (Yadav, 2005). For instance, honey is produced by bees who are nurtured in the forest and their regeneration happen through insect's population. Similarly, hives are nurtured by wildlife through natural process. In economic, the growth of organic agriculture is measured by demand and supply of organic products and inputs.

The area of organic land and production increase through the more farmers adopt organic farming. But in India most of farmers are small and marginal, they are unable to bear the cost burden of certification and accreditation. Although they are interested to join organic farming due to less input cost and they can make their own compost. They are also consensus with natural resources like fertility of land, water quality and food quality. The farmers can earn profit under organic farming through selling products at high premium price. The increasing demand of organic products will help attract farmers towards organic method of cultivation, but there is need for some financial support for small farmers. The new PGS, a cost-effective accreditation and certification method may help farmers to adopt organic farming. The development of organic farming sector will help to create job opportunity for rural and urban young people. The processing, packaging and marketing organic inputs and output are new are for employment. The organic products can be sold both local, national and international market at higher price (Buck et al., 1997). Thus, the organic farming adopted the basic principle that will helps to preserve the natural resources and regenerate the productive capacity of soil and water (Brandt & Mølgaard, 2001; Mäder et al., 2002).

Growth in Certified Cultivated Organic Area

India has the climatic advantage to grow all kinds of crops originally. The diverse climatic condition provide ecosystem for all kind of plants. The indigenous people are nature friendly and living with it. The inherited traditional method of cultivation using domestic livestock and straw based compose will helps to adopt easily organic farming. India have opportunity to explore it by exporting organic products to rest of the world. India having the largest numbers of organic growers in the world.

The organic certified area was recorded 0.002 million hectares in 2000.01, which was further increased 4.72 million hectares in 2013–2014. The certified are having 85 per cent under the category of wildlife and forest collection but in 2018 it was recorded 50 per cent both cultivated and wild collection. As per available data of 2015–2016, India having 5.71-million-hectare data of certified land. The classification was that the 74 per cent area belongs to wildlife and forest collection and only 26 per cent was recorded as cultivated area (APEDA, 2016). But now it changed, and cultivated area increased, and some wildlife and forest were removed form organic category. As per available statistics of 2018, India's rank is 9th in terms of certified organic land in the world (APEDA, 2019; FIBL& IFOAM, 2018).

Table 11.1 shows that the growth of certified organic land producers in India since 2000 to 2013–2014. The rate of growth of certification varied from year to year. The years recorded highest growth rate, in those years government lunch certification schemes. After the completion of these years, farmers quit organic farming. Some

Table 11.1 *Certified Areas and Producers in India (2000–2001 to 2013–2014)*

Year	Total Certified Area (ha)	Share of Ag Land	% Growth of Area	No. Producers	% Growth of Pro
2000–2001	2,775	0.00	–	1426	–
2001–2002	41,000	0.02	1,377.48	5,661	296.98
2002–2003	37,050	0.02	–9.63	5,147	–9.08
2003–2004	76,326	0.04	106.01	5,147	0.00
2004–2005	114,037	0.06	49.41	5,147	0.00
2005–2006	150,790	0.10	32.23	5,147	0.00
2006–2007	528,171	0.24	250.27	44,926	772.86
2007–2008	1,030,311	0.57	95.07	195,741	335.70
2008–2009	1,018,470	0.57	–1.15	340,000	73.70
2009–2010	1,180,000	0.66	15.86	677,275	99.20
2010–2011	780,000	0.43	–33.90	400,551	–40.86
2011–2012	1,084,266	0.60	39.01	547,591	36.71
2012–2013	5,00,000	0.28	–53.89	600,000	9.57
2013–2014	5,10,000	0.29	2.00	650,000	8.33
CV	0.88	–	–	1.11	–

Source: FiBL–IFOAM Various Annual Reports.

year recorded negative growth rate from the previous year, that is, 2002–2003, 2008–2009, 2010–2011 and 2012–2013. The worst performing year was 200–2011 and 2012–2013. These four years growth rate of certified area from previous year was negative, but the numbers of growers are continually increasing except the years 2002–2003 and 2010–2011. In India two third area falls under the category of wildlife and forest collection which is by default organic. Thus, the organic cultivated land is very small less than 1 per cent as compared to conventional land. So, there is need to reduce the cost of certification helps those farmers who want to adopt organic farming. The value of coefficient variation (CV) is 0.8 which shows that there is consistency in the growth of organic farming land and for growers it is 1.1 have some variation.

Table 11.2 presented the status and growth of certified organic land and producers across the years for two time period. The state wise data available from year 2005–2006 to 2013–2014. So, for the purpose study have divided into two group; years (2005–2006 to 2009–2010) for Period-I and years (2019–2010 to 2013–2013) for Period-II. The major states data are shown and in others category fall remaining all states.

The small states like Delhi, Manipur, Tripura and Jammu and Kashmir shows variation the area of cultivation having high value of CV and negative growth in Period-II and positive in Period-I. It was happening because during the year 2008–2009 to 2011–2012 there was increasing trend in certified area. Though, the AAGR for all years recorded negative but leading states recorded positive. At the same time country as whole recorded growth in certified cultivated area. The country as whole AAGR was recorded 7.9 per cent having hope for organic policy makers.

CAUSES OF CLUSTERS IN ORGANIC FARMING IN INDIA

The concept of cluster is really seen in industrial economics. It is widely used across different sectors of the economy for explaining the concentration of industry. Marshall has defined 'the cluster as con-centration of same similar kind of industries in a location tremored as

Table 11.2 State-Wise Total Certified Organic Land (2005–2006 to 2013–2014, ha)

State	Period –I	AAGR	Period II	AAGR	I+II	AAGR	CV
Andhra Pradesh	19,407.85	41.09	13,401.76	–36.21	14,786.8	–6.63	0.75
Arunachal Pradesh	927.06	–24.36	592.80	–183.33	633.5	–134.17	0.99
Assam	4,067.20	19.21	2,866.35	–18.26	3,284.3	–5.38	0.45
Bihar	244.26	50.00	2,424.21	–1,018.12	1,360.7	–669.45	2.48
Chhattisgarh	353.44	–7.12	1,572.94	27.39	1,020.8	16.13	1.38
Delhi	8,236.16	3,040.76	166.39	–4,526.59	4,638.3	–2,427.5	2.61
Goa	9,097.78	–2.28	10,527.20	–4.87	10,081.6	2.17	0.35
Gujarat	59,347.95	–5.89	50,635.25	–5.68	53,264.1	–9.23	0.99
Haryana	6,058.68	–5.09	10,509.42	–42.93	8,207.5	–24.22	0.75
Himachal Pradesh	7,968.34	–619.10	3,414.52	–619.92	6,259.8	–297.2	0.82
J&K	17,787.47	–1,928.44	3,898.98	–342.15	11,979.9	–940.96	1.27
Jharkhard	453.78	24.94	203.45	38.3	365.1	33.29	2.30
Karnataka	30,947.04	2.59	35,688.86	–22.86	31,301.2	–20.42	0.72
Kerala	13,506.20	–3.19	12,569.22	–12.16	12,834.2	–12.46	0.27
Manipur	6,245.24	–25.69	1,454.30	–2,364.75	3,925.2	–1,457.46	1.24
Maharashtra	137,151.96	15.75	119,606.00	–31.8	125,924.7	–3.4	0.65

(Table 11.2 Continued)

(Table 11.2 Continued)

State	Period-I	AAGR	Period II	AAGR	I+II	AAGR	CV
Madhya Pradesh	256,431.83	42.12	267,463.26	-12.19	242,105.6	8.93	0.65
Mizoram	19,203.91	30.58	9,721.98	-137.93	12,974.4	-71.57	1.06
Meghalaya	1,162.64	22.38	1,580.93	-199.17	1,186.1	-121.14	1.02
Nagaland	11,221.28	7.69	5,419.39	-137.26	8,173.1	-64.25	1.00
Orissa	70,790.54	22.02	45,879.21	-60.47	54,178.6	-31.39	0.59
Punjab	3,631.21	-10.8	3,070.51	-97.98	3,138.3	-67.83	0.65
Rajasthan	28,229.84	13.53	56,693.72	8.58	42,610.0	6.83	0.52
Sikkim	2,205.45	-172.84	23,080.07	-68.98	13,225.9	-117.93	1.89
Tripura	60.39	25.00	209.36	-1,652.55	118.6	-1,035.97	1.32
Tamil Nadu	6,666.27	2.72	5,055.07	-35.47	5,762.6	-18.02	0.38
Uttar Pradesh	21,314.33	47.32	39,611.24	12.92	27,898.1	18.51	0.67
Uttarakhand	17,730.88	30.81	23,391.40	-19.92	19,394.0	-1.12	0.57
West Bengal	11,289.34	17.32	8,831.85	-282.13	9,449.2	-171.31	0.71
Others	2,237.75	36.63	625.97	41.98	1,591.0	39.98	1.93
India	773976.11	30.67	760,165.59	-7.95	731,673.3	7.93	0.47

Source: National Centre of Organic Farming, Various Annual Reports.

cluster' in his book *Principle of Economics* published in 1890. Porter in his book *The Comparative Advantage of Nations* in 1990 has defined as 'the group of suppliers and firms in specialized and related industries are concentrated in specific place. His though was limited to vertical integration and value chain. Further in 1998, Porter has primarily emphasized the comparative advantage of cluster. He has found that cluster have three type advantages; first is increasing productivity of firm which are in cluster. Second is increasing rate of innovation in cluster that help to future industrial growth. Third is inspiration to start new business in the cluster. In 1996 Baptista pointed out the technological innovation is the key for growth of cluster and attract new firms to start business. Latter in 1999 Baptista and Swam has examine the role of geographical concentrating in technological innovation and organizational development.

It became popular is the industrial economics after the porter validation and applied for research and policy formation. The cluster is defined by various economist within the framework of geographical concentration of interrelated firms and industries. The benefit of this concentration is that increase the competition and pressure among the industries. The approach of cluster is applied by entrepreneurs and for study the potential advantage of regional growth and competition (Smith, 2003). Andrew (2003) has claimed in his work that the cluster analysis helps policy makers to identify local concentration of firms and industries based on employment. He found that employment opportunity and availability of resources attract inverters and future growth of cluster.

The three essential condition are identified by Banner (2004) for emergence of local industrial cluster. The *first* condition is the region should be factor endowment for related industry. The availability of relevant factors, that is, industrial structure, qualified labour, geographical location, basic infrastructure and policy support, is called the *pre-requisites* of arising a cluster. The *second* condition is essential and favourable conditions such as; specific innovation, policy support, promoting activity, historical events and support for leading firms are called *triggering events*. The third condition is along with the above two necessary condition the *self-augmenting process* is complimentary for emergence of local industrial cluster. The self-augmenting process includes; cooperation among the firms, supplier-buyers relation, freedom to choose of co-location, accumulation of labour force, intra and inter industrial spillovers, start-ups and cooperate with local society. These activities help to make environment for industrial development (Brenner & Muhlig, 2007).

The conditions which are discussed above are for the formation of local and regional cluster. In case of organic farming the conditions may be different but theoretically similar conditions require to develop cluster. The condition which require for the developed a region as organic farming cluster are, education of farmers, farm size, training of farmers, facility of accreditation and certification, livestock, facility of organic input and output market and public policy. In India states like Himachal Pradesh, Madhya Pradesh, Maharashtra, Karnataka, Rajasthan, Uttar Pradesh and Uttarakhand have developed some area as cluster of organic. These area or districts concreted with organic farming due to central and state sponsored polices and schemes on organic farming along with the climatic conditions. The rainfed districts and regions are promoted for the organic farming and government will help to develop as organic area.

The concept of industrial cluster is applied by Naik and Nagadevarais (2010) for the study of organic farming in Karnataka. They have focused on the study of organic pulses growers cluster in Karnataka using cost benefit analysis. The have found that in case of cluster villages the cost of organic cultivation was lesser than non-clutter villages. They found that were the cluster created for organic cultivation, the organic input production unit are open by the inverters and output market also developed. Thus, the cluster will increase the efficiency of producer and increase the productivity and income of the producers.

Thus, the growth of organic farming as cluster requires some condition as mentioned above. These factors may be classified into three categories as, institutional requirement, climatic conditions, production factors. The category first, includes institutional included, public policy, legal laws and regulation, accreditation and certification agencies, financial institutions, training facility, market facility, etc. The category second, includes rain fall condition, soil type, weather and climate and socio-economic conditions. The category third includes, farm size, size of livestock, farmers education, family work force, self-made inputs, organic seeds, etc. These following steps can help to high growth of organic farming and formation in clusters. (a) The accreditation and certification by government agency at low cost can help in creation of organic cluster. (b) The facilities of organic inputs out market played crucial role in creation organic growers cluster. (c) The climatic condition, that is, types of soil, crop patter, irrigation facility and organic inputs are necessary for creation of organic cluster. (d) The financial accessibility to producer at lower rate of interest will help farmers to make initial expenses on organic farming. (e) The learning by doing is method of increasing production which create cooperation among the producers in the village and area. (f) The Participatory Guarantee System (PGS) of certification will help to reduce cost of accreditation and creation of cluster. (g) The health concern of people and ecosystem will help is creation of organic cluster.

CONCLUSION

India having the largest number, 6.5 million, of organic producers in the world, but the cultivation area under organic farming is still below 2 per cent of total cultivated land. The certified area under organic farming was only 41,000 hectares, constituting just 0.03 per cent area of the total cultivated area, in the year 2002. It was recorded as 5.71 million ha organic certified land in 2015–2016. But this certified area still has a share of only of 26 per cent (1.49 million ha) in cultivation and remaining 74 per cent (4.22 million ha) is under forest and wild collection of minor forest produces. Sot the country has very negligible certified land under organic cultivation and this can be increased manifold by promoting organic farming in rainfed regions. The growth of organic farming is not similar across India; it varies farm state to state. The states like Madhya Pradesh, Gujarat, Andhra Pradesh, Rajasthan, Maharashtra, Uttar Pradesh and North eastern states are leading states in organic production.

REFERENCES

Bacon, C. (2005). Confronting the coffee crisis: Can fair trade, organic, and specialty coffees reduce small-scale farmer vulnerability in northern Nicaragua? *World Development*, *33*(3), 497–511.

Baptista, R. (1996). *An empirical study of innovation, entry and diffusion in industrial clusters* (Unpublished doctoral thesis). London Business School, London.

Baptista, R., & Swann, G. P. (1999). A comparison of clustering dynamics in the US and UK computer industries. *Journal of Evolutionary Economics*, *9*(3), 373–399.

Barker, V. A. (2010). *Science and technology of organic farming.* London and New York, NY: CRC Press.

Beyl, C. A. (1992). Rachel Carson, silent spring, and the environmental movement. *HortTechnology*, *2*(2), 272–275.

Brandt, K., & Mølgaard, J. P. (2001). Organic agriculture: Does it enhance or reduce the nutritional value of plant foods? *Journal of the Science of Food and Agriculture*, *81*(9), 924–931.

Brenner, T. (2004). Local industrial clusters: Existence, emergence and evolution. London: Routledge.

Brenner, T., & Mühlig, A. (2007). *Factors and mechanisms causing the emergence of local industrial clusters: A meta-study of 159 cases* (Papers on economics and evolution no. 0723). Retrieved from http://www.evoecon.mpg.de/fileadmin/user_upload/Paper/2007-23.pdf

Buck, D., Getz, C., & Guthman, J. (1997). From farm to table: The organic vegetable commodity chain of Northern California. *Sociologia Ruralis*, *37*(1), 3–20.

Haedicke, M. A. (2012). Keeping our mission, changing our system: Translation and organizational change in natural foods co-ops. *The Sociological Quarterly*, *53*(1), 44–67.

Herr, A. (2003, 14 October). Industry cluster analysis of Westmoreland and Fayette counties addressing workforce development needs. Prepared for the Westmoreland-Fayette Workforce, 44.

Kuepper, G. (2010). *A brief overview of the history and philosophy of organic agriculture.* Poteau: Kerr Center for Sustainable Agriculture.

Lægreid, M., Bockman, O. C., & Kaarstad, O. (1999). *Agriculture, fertilizers and the environment.* Wallingford: CABI Publishing.

Mäder, P., Fliessbach, A., Dubois, D., Gunst, L., Fried, P., & Niggli, U. (2002). Soil fertility and biodiversity in organic farming. *Science*, *296*(5573), 1694–1697.

Marwaha, B.C. (2005). Is India in a position to switch over to pure organic farming in totality? *Indian Journal of Fertilizers*, *1*(7), 47–52.

Naik, G., & Nagadevara, V. (2010). Spatial clusters in organic farming—a case study of pulses cultivation in Karnataka (Working Paper No. 316). Bangalore: IIM Bangalore.

Narayanan, S. (2005). *Organic farming in India: Relevance, problems and constraints.* Mumbai: National Bank for Agriculture and Rural Development.

Niggli, U., & Lockeretz, W. (1996). Development of research in organic agriculture. In Oestergaard, T. (Ed.), *Fundamentals of organic agriculture* (pp. 9–23). Tholey-Theley IFOAM.

Porter, M. E. (1990). The competitive advantage of notions. *Harvard Business Review, 68*(2), 73–93.

Porter, M. E., & Porter, M. P. (1998). Location, clusters, and the 'new' microeconomics of competition. *Business Economics, 33*(1), 7–13.

Porter, M. E. (1998, November–December). Clusters and the new economics of competition. *Harvard Business Review, 76*(6), 77–90.

Prasad, R. (2005). Organic farming vis-à-vis modern agriculture. *Current Science, 89*(2), 252–254.

Shukla, U. N., Mishra, M. L., & Bairwa, K. C. (2013). Organic farming: Current status in India. *Popular Kheti—Special Issue on Organic Farming, 1*(4), 19–25.

Smith, R. V. (2003). *Industry clusters analysis: inspiring a common strategy for community development* (p. 1). Lewisburg, PA: Central Pennsylvania Workforce Development Corporation.

Venkateswarlu, B., Balloli, S. S., & Ramakrishna, Y. S. (2008). *Organic farming in rainfed agriculture: Opportunities and constraints* (p. 185). Hyderabad: Central Research Institute for Dryland Agriculture.

Yadav, V., & Malanson, G. (2008). Spatially explicit historical land use land cover and soil organic carbon transformations in Southern Illinois. *Agriculture, Ecosystems & Environment, 123*(4), 280–292.

Health and Environmental Concerns

INTRODUCTION

Health and education are the true measures of human development, but access to these two needs income. In the current era of market-based system, income is crucial for human development because both health and education are becoming more marketable. Health is crucial because it directly affects the outcomes and performance of people. Proper diet, clean water and fresh air are crucial for good health. All these three intakes are polluted by human activity more than natural disasters by the use of chemical fertilizers and pesticides in agriculture production. There are various evidences in the world that the use of chemical fertilizers and pesticides has an adverse impact on human health and the environment. The use of chemical fertilizers and pesticides in agriculture production activity has serious impacts on public health and the environment in United States (Pimentel et al., 2005). For example, in United States more than 90 per cent of corn farmers use atrazine pesticide for weed control and it is found in streams and groundwater (Pimentel et al., 1993; USGS, 2001). The estimated environmental and health care cost of pesticide use at recommended levels in the United States is about $12 billion every year (Pimentel, 2005).

This chapter presents the experimental findings of conventional and organic farming on the health-promoting qualities of soil, water and other environmental components, along with the evidence on human health on conventional and organic food consumption. In the literature both types of results are found—those which support the idea that organic foods have more nutritional value and others whose findings do not support the above (Forman & Silverstein, 2012).

What are the influencing factors for consumers to buy organic food is also discussed. In the literature it is widely found that health and environment both are crucial factors affecting the choice of organic food to consumers. The general belief of consumers is that organic produce is more nutritious than conventional food, but the research and finding do not support this unanimously. So, more research is essential to prove the difference in nutritional value of organic food and conventional food.

HEALTH AND NUTRITION ISSUES OF ORGANIC PRODUCT

The literature on nutritional ingredients on organic and non-organic products has another line of debate in academics. The nutritional ingredients depend on various factors such as soil quality, seeds, fertilizer, water quality and ecosystem. The empirical work conducted by Albert Howard in 1947 motivated him to set up an association for soil management. The aim of this association was to adopt an integrated approach for soil, plants, human and animals' management. At the same time the quality of food differentiation under organic farming was challenged. Howard and Balfour have rejected the concept of increasing yields through using chemical fertilizer and HYV seeds. With their own experiences, they found that conventional farming degraded the soil's ingredients and fertility. This decline of soil fertility was due to monocropping and using the same chemical fertilizers (Heaton, 2001).

Another scholar in this debate was Worthington (1998), who reviewed the literature on this nutritional and quality debate on both inputs and outputs. He found that nutritional content of different cultivation method are different, but for some products it is similar. In the case of carrot, zinc and broccoli, vitamin C was recorded 40 per cent higher in organic products. Thus, concluded that the nutritional content was higher in organic cultivation of some products. On the same lines he further conducted another study in 2001 for some agriculture products and found magnesium, phosphorous, vitamin C and iron recorded more in organic products than conventional products. But in the same study he also recorded that organic products had lower amounts of heavy metal, nitrates and proteins (Worthington, 2001).

Another work on nutritional and quality issue was conducted by Saba and Messina in 2003. They conducted an empirical study based on consumer response for food grains, fruits and vegetables in both types of consumers, organic and conventional, based on 947 respondents in Italy. The cluster group method was applied for segregation of both types of consumers and they found that the consumers of conventional products had less positive attitude towards organic products than those facing heath issues. The other group, those facing heath issues, had strong attitude towards organic products (Sparks & Shepherd, 1992).

On the same lines Lombardi-Boccia et al. (2004) conducted scientific empirical study for yellow plum cultivated in both type of farms and found minimum nutritional differences in both types of products. The only difference was recorded in phenolic, vitamins, C and E, and β-carotene. Only antioxidants recorded nutritional benefits in organic products. He also suggested that organic products have more antioxidants to protect from fungus and insects' attacks. But Caris-Veyrat et al. (2004) conducted a similar study and did not record much differences in antioxidants in the two types of products.

Another study which was conducted by Rekha et al. (2006) in five states namely Uttar Pradesh, Haryana, Maharashtra, Chhattisgarh and Uttarakhand for rice and wheat. The sample of both type of food grains was collected for the market and pesticides contents were studied in a laboratory. For both rice and wheat in organic products no pesticides and fertilizers evidence was found, but in the case of conventional wheat and rice, these contents were present. Further scientific study was conducted by Rist et al. (2007) on the breast milk of mothers who consumed organic and non-organic food items. They found that rumenic acid, which is an isomer of conjugated linolenic acid and trans-vaccenic acid, was high in organic food item-consuming mothers.

The nutritional estimate of both types of products was conducted by Hoefkens et al. (2010) in adult age groups for vegetables. They used probability simulation approach to know the consumers' response. They divided into two adults' groups, n = 3,245 from Belgians and n = 552 from Flemish organic consumers. The results showed significant differences in nutritional strength in both types of vegetables.

The organic vegetable consumers outweighed conventional consumers. The meta-analysis was conducted by Palupi et al. (2012) for both types of dairy products and studied nutrients such as proteins, trans-11conjugated acid, total omega-3 fatty acid, trans-11 vaccenic acid, eicosapentaenoic acid and docosapentanoic acid. They recorded higher amount of these nutrients in organic dairy products than in conventional dairy products.

The heath consciousness for organic products was studied by Batra et al. (2014) who found that consumers are ready to pay higher for organic product only due to higher nutrients and their own health consciousness. They also made nutritional ingredients comparison for both types of products available in the market and recorded no significant differences in calorie and proteins, but fat content in egg and baby food were found less in organic products. On the same lines, the *Science Review* (2014) have produced its result on nutritional comparison for organic and non-organic food items. This review has found that organic food items recorded higher quantity of potassium, nitrates, phenolics, antioxidant, vitamin E, A and C than conventional products. A similar study was conducted for wheat flour available in the market by Vreek et al. (2014). They found lower amount of protein, calcium, manganese and iron in organic floor, but the digestible proteins, zinc, molybdenum and potassium were found higher in amount.

ENVIRONMENTAL CONCERN OF ORGANIC FOOD

The empirical studies presented here were conducted to address known consumers' and producers' concerns on organic farming in terms of environmental conservation. In this regard Bulluck III et al. (2002) conducted a study based on sacrifice experiment through testing the soil of three farms each of the two kinds in Maryland and Virginia in 1996 and 1997. In the farms, same variables are grown using organic and conventional method of cultivation. They recorded higher density of microorganisms such as thermophilic, trichoderma and bacteria and lower density of pythium and phytophthora in organic farms, while higher magnesium, potassium, manganese and calcium were found in organic soil than conventional soil.

Another study was conducted by Liua et al. (2007) based on the analysis of soil ingredients in both types of farms. The soil sample was taken from 10 field locations of three methods of cultivation. Their testing-based finding shows that both sustainable and organic farms improved all types of ingredients as compared to conventional farms. A similar study was conducted by Fließbach et al. (2007) in 1978 for three types of farming namely, biodynamic, bio-organic and integrated method. They have recorded intensity of fertilizer using 0.7 and 1.4 livestock units for a period of seven years. Crop rotation was done and after seven years three consecutive experiments showed that bio-organic farms had the most fertile soil. In bio-organic farms both biological and organic components' quantities depended on the used types of fertilizer.

On the same lines a 30-year long-term experiment-based study was conducted by Parvathi et al. (2013) for cultivation of groundnut in Alfisol. Both organic and conventional farms were taken for the experiment for intensive cropping of groundnut. They used 20:10:25 NPK per hectare along with zinc sulphate and gypsum at conventional farm recorded 1,499 kg per hectare higher yield than NPK plus FYM along with gypsum applied in organic farm. But the EC and pH value of soil did not change in both farms. Thus, they concluded that balanced application of fertile improves the fertility of both type of soil.

A similar field experiment-based study was conducted by Singh (2013) at Delhi, India in 2010 and 2011. He studied the use of water efficiency in rice cultivation in three farming methods, namely, integrated, organic and non-organic. He conducted nine treatment using different nutrients supplements using three type of farming, namely, aerobic rice system (ARS), system of rice intensification (SRI) and conventional transplanting (CT). He found that SRI and CT recorded higher yield than ARS, but the use of water efficiency was higher—34.5 to 36 per cent higher in SRI and 28.9 to 32.1 per cent higher in ARS than CT.

Another study was conducted by Larsen et al. (2014) for knowing biological components of soil under both conventional and organic corn growing farms in Mill River and Mountain region of North Carolina in United States. The study was conducted during 2010 to 2012 and he has found that with organic no till treatments having twice TC quantity, Light Fraction Particulate Organic Matter (LPOM) in the upper 15cm and four times the quantity of Microbial Biomass Carbon (MBC) as compare to conventionally tilled treatments. It was happening in conventional farm due to the use of hybrid crops but was reducing in organic farm due to lack of weed control.

The effect of NPK and organic manure was studied by Zhang et al. (2014) for the period from 1989 to 2009 in China. They have conducted long term experiment and made septation of soil into five categories: clay fraction (less than 2 µm), silt (2–53 µm), small (53–250 µm), medium (250–2000 µm) and large macro-aggregates (2000 µm). They have measured phospholipids fatty acid and carbon concentration and found that it increased 123 to 134 per cent by using manure, but decreased oxygen diffusion coefficients. But using NPK increases only organic carbon in macro aggregates and combining both decreases M/B ratio in all aggregates except silt. The organic carbon is negatively correlated with use of manure and NPK along with M/B ratio with having ($P < 0.05$).

An experiment-based study was conducted by Srinivasaro et al. (2014) using chemical fertilizer and manure for knowing crop productively and C sequestration in duration of 18 years. They took castor, cluster bean and pearl millet in western India. The result shows that the net loss of soil organic carbon depletion was C 4.4 mg/ha organic farm while C 12 mg/ha in controlled experiment in 18 years.

Similar field experiment-based study was conducted by Dubey et al. (2014) for soil content and water use in different rice cultivations in Jabalpur MP India during 2004 to 2008. They conducted 12 trail treatments using three nutrients: 100 per cent non-organic, 100 per cent organic, integrated with 50 per cent organic and 50 per cent non-organic for four crop cycles. The productivity was highest 187.16 quintal per hectare with having water use 86.5 kg per hectare per cm in Parbhani Kranti cycle as compared to other cycles.

CONCLUSION

Many studies have confirmed that the growing demand for organic products has two major underlying concerns: one is health and another is environmental. The health concerns are related to nutritional enrichment of organic products, along with their being pesticide and fertilizer free. The environment concerns of consumers are related to saving environmental domains from degradation and pollution via modern conventional agriculture practices. Both concerns of consumers lead to increase in the demand for organic products and the resultant producers' wish to produce organically.

REFERENCES

US Geological Survey (USGS). (2001). *Selected findings and current perspectives on urban and agricultural water quality by the national water-quality assessment program*. Washington, DC: US Department of the Interior, USGS.

Batra, P., Sharma, N., & Gupta, P. (2014). Organic foods for children: Health or hype? *Indian Pediatrics, 51*(5), 349–353.

Bulluck, L.R., III, Brosius, M., Evanylo, G. K., & Ristaino, J. B. (2002). Organic and synthetic fertility amendments influence soil microbial, physical and chemical properties on organic and conventional farms. *Applied Soil Ecology, 19*(2), 147–160.

Dubey, R., Sharma, R. S., & Dubey, D. P. (2014). Effect of organic, inorganic and integrated nutrient management on crop productivity, water productivity and soil properties under various rice-based cropping systems in Madhya Pradesh, India. *International Journal of Current Microbiology and Applied Sciences, 3*(2), 381–389.

Heaton, S. (2001). *Organic farming, food quality and human health*. Bristol: Soil Association.

Hoefkens, C., Sioen, I., Baert, K., De Meulenaer, B., De Henauw, S., Vandekinderen, I., ... & Van Camp, J. (2010). Consuming organic versus conventional vegetables: The effect on nutrient and contaminant intakes. *Food and Chemical Toxicology, 48*(11), 3058–3066.

Larsen, E., Grossman, J., Edgell, J., Hoyt, G., Osmond, D., & Hu, S. (2014). Soil biological properties, soil losses and corn yield in long-term organic and conventional farming systems. *Soil and Tillage Research, 139*, 37–45.

Liu, B., Tu, C., Hu, S., Gumpertz, M., & Ristaino, J. B. (2007). Effect of organic, sustainable, and conventional management strategies in grower fields on soil physical, chemical, and biological factors and the incidence of Southern blight. *Applied Soil Ecology, 37*(3), 202–214.

Lombardi-Boccia, G., Lucarini, M., Lanzi, S., Aguzzi, A., & Cappelloni, M. (2004). Nutrients and antioxidant molecules in yellow plums (*Prunus domestica L.*) from conventional and organic productions: a comparative study. *Journal of Agricultural and Food Chemistry, 52*(1), 90–94.

Palupi, E., Jayanegara, A., Ploeger, A., & Kahl, J. (2012). Comparison of nutritional quality between conventional and organic dairy products: A meta-analysis. *Journal of the Science of Food and Agriculture, 92*(14), 2774–2781.

Parvathi, E., Venkaiah, K., Munaswamy, V., Naidu, M. V. S., Giridhara Krishna, T., & Prasad, T. N. V. K. V. (2013). Long-term effect of manure and fertilizers on the physical and chemical properties of an alfisol under semi-arid rainfed conditions. *International Journal of Agricultural Sciences, 3*(4), 500–505.

Pimentel, D., Hepperly, P., Hanson, J., Siedel, R., & Douds, D. (2005). Organic and conventional farming systems: Environmental and economic issues. Technical Report No 05-1, Cornell University, Ithaca: NY, USA.

Pimentel, D., McLaughlin, L., Zepp, A., Lakitan, B., Kraus, T., Kleinman, P., …, Selig, G. (1993). Environmental and economic effects of reducing pesticide use in agriculture. *Agriculture, Ecosystems and Environment, 46*(1–4): 273–288.

Rist, L., Mueller, A., Barthel, C., Snijders, B., Jansen, M., Simoes-Wüst, A. P., … & Thijs, C. (2007). Influence of organic diet on the amount of conjugated linoleic acids in breast milk of lactating women in the Netherlands. *British Journal of Nutrition, 97*(4), 735–743.

Saba, A., & Messina, F. (2003). Attitudes towards organic foods and risk/benefit perception associated with pesticides. *Food Quality and Preference, 14*(8), 637–645.

Singh, Y. V. (2013). Crop and water productivity as influenced by rice cultivation methods under organic and inorganic sources of nutrient supply. *Paddy and Water Environment, 11*(1–4), 531–542.

Sparks, P., & Shepherd, R. (1992). Self-identity and the theory of planned behavior: Assessing the role of identification with 'green consumerism'. *Social Psychology Quarterly, 55*(4), 388–399.

Srinivasarao, C. H., Venkateswarlu, B., Lal, R., Singh, A. K., Kundu, S., Vittal, K. P. R., … & Patel, M. M. (2014). Long-term manuring and fertilizer effects on depletion of soil organic carbon stocks under pearl millet-cluster bean-castor rotation in western India. *Land Degradation & Development, 25*(2), 173–183.

Worthington, V. (1998). Effect of agricultural methods on nutritional quality: A comparison of organic with conventional crops. *Alternative Therapies in Health and Medicine, 4*(1), 58–69.

Worthington, V. (2001). Nutritional quality of organic versus conventional fruits, vegetables, and grains. *The Journal of Alternative & Complementary Medicine, 7*(2), 161–173.

Zhang, H., Ding, W., He, X., Yu, H., Fan, J., & Liu, D. (2014). Influence of 20–year organic and inorganic fertilization on organic carbon accumulation and microbial community structure of aggregates in an intensively cultivated sandy loam soil. *PLoS One, 9*(3), e92733.

Public Policy

International Policy and Regulation

INTRODUCTION

People are living in the midst of multiple crises all over the world. Healthy food and clean water are the basic requirements of humanity. But many countries are unable to provide these two necessities. The agriculture sector has made tremendous progress and increased productivity many times and millions of people have been removed from poverty. At the same time the modern cultivation method creates many obstacles to achieve agricultural sustainability. In the 21st century the major changes among the countries of the world concern how to make production and consumption activities sustainable. For achieving overall sustainability, each production and consumption activity must be sustainable. How to achieve sustainability in agricultural activity is the most debated issue in the current century. The whole debate is on the method of cultivation under which minimum damage to natural resources and maximum output can be attained.

Organic farming is getting popular across the countries since the end of the 20th century. Some international organizations have initiated the spread of information about the environmental and health benefits of organic farming. At the same time the quality and acceptability of the products cultivated under organic method of production are under assessment. This is certainly a major challenge for national and international organizations that are engaged in promotion of organic farming. The movement was started by IFOAM and the Research Institute of Organic Agriculture (FiBL) in the 1970s. So, the product quality check and inspection of farming method are possible through only a registered body that has acceptance among the consumers. These bodies are registered with government and get authority to certify the land and method of cultivation. These certification bodies are called 'third party certification systems' and are most popular across the counties. But the cost of certification and accreditation is very high under third party certification system. Now various low-cost certification methods such as those used by government certification agencies, agriculture departments, group certifications and so on have been initiated.

This chapter critically evaluates the international principles, policies and regulations for organic farming. The international rules and regulations for certification are discussed briefly. International standards have been set up for use of organic inputs, certification of land, packaging, organic seeds and other related activities. All these activities are directly or indirectly related to cultivation of organic products. The basic four principles, under which the organic farming practices are examined, are mostly related to logical and ethical reasoning.

INTERNATIONAL POLICIES AND REGULATIONS

The organic movement began with the value judgement of organic food and environmental concerns. Those who had their own opinion on organic farming had a different school of thought. The one group of people who supports the organic farming movement have optimistic views on organic farming. Although they know that under this method of cultivation productivity declines, but at the same time this method supports the environmental cause. While, another group of people who oppose organic farming have raised several questions on the reliability of this farming method. They argue that if growth population is higher than the growth of food grains, then millions of people will suffer from the hunger and poverty. A group of people who follow the middle path also support this view to some extent. They have suggested that organic and conventional farming go together. One feeds the demand of organic products and other feeds the demand of masses. This approach is more realistic because initially most countries were unable to grow organic food for the masses.

The organic movement stared with the awareness programme of both consumers and producers. During the 1970s IFOAM took initiatives for promotion of organic movement in United States, Europe and South Africa. This organization's goal was to provide a platform to exchange information with its members, and later share its information with like-minded people those that are working for promotion of organic farming in other regions of the world. It provides all kinds of support to producers, distributors, consumers, private and government bodies (Niggli & Lockeretz, 1996). It also facilitates certification and regulations to the member nations along with exchange of information. It is continuously working for harmonization of rules and regulation. The harmonization of rules and regulation helps in increasing the world trade of organic products. However, IFOAM has made progress to make common consensus across the counties of the world on the certification system known as the third party certification system.

The formation of private third-party standards on organic farming began in the 1980s. The IFOAM standards on organic farming are known as 'standards of standards' and are generally called IFOAM Basic Standards (IBS). There is a provision that verification agencies can set their own standards within the IBS framework. The International Organic Accreditation Services (IOAS) is another platform, where agencies and countries can apply for IFOAM accreditation. It monitors through the annual surveillance management system, which includes annual visits of office bearers (Anon 2005a).

Presently, more than 132 regulations on organic agriculture are in practice across the world. A joint initiative has been taken by IFOAM, FAO and UNCTAD for harmonization of organic agriculture regulations and standards across the nations. The UNDP has accepted that organic agriculture helps attain Millennium Development Goals (MDG) under Agenda 21. IFOAM was established in 1972. Now, over its 45 years of journey, it has more than 800 member organizations affiliated from 180 countries across the world. It provides membership to all types of organic agents such as nations, farmer's cooperative societies, certification agencies, consumer organizations and consulting agencies of trade (Morgera et al., 2012; Luttikholt, 2007).

IFOAM provides the platform to organic producers, consumers, accreditation agencies and countries to share information on organic agriculture with each other and bridge the knowledge gap between them. It laid down the foundations for growth of organic agriculture across the world. It provided a global forum for opinion, discussion and reflection on issues such as how an organization can claim that it has expressed the value of organic movement with diverse membership system and how these values have been institutionalized and implemented for the whole organic movement. These led to the birth of principles of organic farming.

PRINCIPLES OF ORGANIC AGRICULTURE

Organic agriculture is based on four basic principles: The principle of health, principle of ecology, principle of fairness and principle of care (IFOAM, 2005a). The Preamble of organic agriculture (IFOAM, 2005a; Luttikholt, 2007) tells about the aim and vision within the framework of these four basic principles to achieve global sustainability. It points out the dependency of humankind on agricultural activity since the inception of human society. The aim of all these principles is to help farmers grow crops without putting the ecosystem and biodiversity under threat. The existence of humankind depends on their quality of life and the quality of natural resources as well as ecosystem. The Preamble talks about both the right and the responsibility of humankind to make agriculture sustainable. It also talks in detail about the standards and provisions for organic agriculture.

The Principle of Health

'Organic Agriculture should sustain and enhance the health of soil, plant, animal, human and planet as one and indivisible.' (IFOAM, 2005a)

This principle tells us that the health of ecosystems is integrated with the health of the community and the individual. The health of ecosystems could be expressed as healthy soil producing healthy crops that foster the health of people and animals.

Here health lies in the integrity and wholeness of living systems. It does not simply mean the absence of illness but the maintenance of mental, physical, ecological and social well-being. The key characteristics of health that one should attain are immunity, resilience and regeneration.

The role of organic agriculture is to enhance and sustain the health of the ecosystems, whether in farming, processing, consumption or distribution. It is intended to produce organically high quality and nutritious food that contributes to health care and well-being. To attain the above, farming should be free of fertilizer, pesticide and animal drugs—those which have adverse effects on health (Luttikholt, 2007).

The Principle of Ecology

Organic Agriculture should be based on living ecological systems and cycles, work with them, emulate them and help sustain them. (IFOAM, 2005a)

The roots of this principle place organic agriculture within the living ecosystem. It reveals that production of crops should be based on ecological processes and recycling. The nutritional quality of food and human well-being are achieved through ecology-friendly production systems, for instance, crops, living soil, animals, farm ecosystems, fish and marine organisms, and the aquatic environment.

In organic farming, wild and pastoral harvest systems ought to fit the cycles and ecological balances in nature. These cycles are universal, but their operation is site-specific. Thus, organic farming must be as per local conditions of ecology, culture and scale. The inputs should be reduced, reused and recycled in such an efficient manner that conserves resources and energy. Finally, organic agriculture must attain ecological balance through the best design of farming systems (ibid., 2007).

The Principle of Fairness

Organic Agriculture should build on relationships that ensure fairness with regard to the common environment and life opportunities. (IFOAM, 2005a)

The term 'fairness' is characterized as equity, respect, justice and stewardship in the collective world, both among people and in their relations with other living beings.

The principle of fairness emphasizes the cooperation and fairness among all stakeholders such as producers, consumers, traders, processors and distributors involved in organic agriculture. Its aim is to produce sufficient supply of quality food, a good life and food sovereignty to everyone involved in the production practices.

The inherent feature of this principle is that animals must be provided opportunities and conditions as per their physiological need, natural behaviour and well-being. The natural and environmental resources which are used in production and consumption must be managed socially, ecologically, inter-generationally and sustainably. Fairness requires production, trade and distribution systems that are open and place an equitable amount of burden in terms of social and environmental costs.

The Principle of Care

Organic Agriculture should be managed in a precautionary and responsible manner to protect the health and well-being of current and future generations and the environment.(IFOAM, 2005a)

Organic agriculture is a living and dynamic method of cultivation which incorporates the condition of both internal and external demand. In this method, practitioners can enhance productivity through efficient utilization of resources, but not at the cost of jeopardizing health and well-being. Subsequently, before adopting new technologies, one must take care of the ecosystem. The existing one must be assessed and reviewed thoroughly first.

The principle of care states that responsibility and precaution are the two main concerns in the development, management and choice of appropriate technology for organic farming. It must be scientifically proved that organic agriculture is healthy, safe and ecologically sound. But only scientific validity in not enough as it should be practically liable and based on indigenous knowledge that offers valid solutions. It should reduce the risk of adopting appropriate technologies and chance of rejection through genetic engineering. The decisions are reflected by the needs and values of all those who might be affected in the process of participation (ibid., 2007).

IMPLEMENTATION OF THE PRINCIPLES

Making rules and regulations is easier than implementation in their true spirit. The IFOAM translated these principles in more than 10 languages such as Spanish, Danish and Chinese for easy accessibility and implementation in December 2006. Organizations like the IFOAM are dealing with only the ethical part of these principles, the legal part is left for the respective countries. It should be ideal that all stakeholders participate in the decision-making process when the organization is dealing with ethical parts of principles. The implementation of principles is an ongoing process and the principles should be circulated and translated based on the specific situation (ibid., 2007).

The implementation of certification and accreditation norms must be as per international standards for universal acceptances of organic products. Third party certification system is easily accessible in all counties, but it is costlier than other systems. For export purpose, the product must be certified under international norms. But for domestic consumers it becomes many times costlier than conventional products. The IFOAM has introduced the PGS, a low-cost certification system under the informal organic agriculture framework. This will provide verification and inspection power to test organic production methods locally thorough community involvement. The PGS group is created by the farmers and the group members encourage direct participation of other farmers and consumers. The belief and acceptance of organic products are popularizing through consumers' presence at farm. The PGS norms are based on four basic principle of organic agriculture along with national rules and regulation. The national rules and regulations are linked with international norms, so it is basically the decentralization of international standards. In PGS system, farmers can explain to consumers and community how their farming complies with these principles (IFOAM, 2008).

The concept of organic sustainability is articulated by the Organic Revision Project, which is very close to the notion of 'ecological justice' (Alrøe, Byrne, & Glover, 2006). This notion overlaps with the principles of organic agriculture. Ecological justice is the fair distribution of all good or bad, along with extremities, among all the living beings on Earth. It is recognized that environments have value greater than the humans. The interdependency of ecological system and human beings is established by this notion. It extends the principle of fairness for all living beings including plants and insects. Thus, this attitude where humans and nature are integrated into each other is the true implication of organic farming. These principles are declared by IFOAM so that all practitioners (producers and consumers) are aware that they have both duties and responsibilities to make organic farming sustainable. It is important for new organic farmers that they learn these principles (ibid., 2007),

CONCLUSION

Rules and regulations are essential to increase the faith of consumers in organic products. The whole world accepts the four basic principles: The principle of health, principle of ecology, principle of fairness and principle of care of organic farming. These principles present the aim of organic farming; for promoting these principles, each nation has developed rules and regulations for producers to maintain quality and produce using only organic manure. Certification and inspection are compulsory for all organic producers to maintain quality of organic products.

REFERENCES

Alrøe, H. F., Byrne, J., & Glover, L. (2006). Organic agriculture and ecological justice: Ethics and practice. *Global Development of Organic Agriculture: Challenges and Prospects Wallingford: CAB International*, 75–112.

Anon (2005a). The IFOAM norms for organic production and processing (version 2005). International Federation of Organic Agriculture Movements (IFOAM), Bonn, p. 132.

International Federation of Organic Agriculture Movements (IFOAM). (2005a). *The IFOAM norms for organic production and processing* (version 2005, pp. 1–121). Bonn: Author. Retrieved from https://www.ifoam.bio/sites/default/files/page/files/norms_eng_v4_20090113.pdf

Luttikholt, L. W. (2007). Principles of organic agriculture as formulated by the International Federation of Organic Agriculture Movements. *NJAS-Wageningen Journal of Life Sciences, 54*(4), 347–360.

Morgera, E., Caro, C. B., & Durán, G. M. (2012). Organic agriculture and the law. FAO Legislative Study (107).

Niggli, U., & Lockeretz, W. (1996). Development of research in organic agriculture. In T. Oestergaard (Ed.), *Fundamentals of Organic Agriculture* (pp. 1–23). Tholey-Theley: IFOAM.

National Policy and Regulation

INTRODUCTION

Since the western whole world is getting momentum towards the organic farming movement during 1990s, India started late in 2000s, when the central government introduced the NPOP in some states. It was the beginning of policy-level consideration on organic farming. The policy document was published by the Ministry of Commerce and Industries in 2005. This document accepted the four basic universally accepted principles: the principle of health, principle of ecology, principle of fairness and principle of care for organic farming and followed the international norms of certification process. The document brought out clear guidelines for organic producers for converting land into organic farm and the process of certification by either through government agency or other existing private agencies. This chapter provides the detailed policy provisions and guidelines for certification agencies and producers.

The NPOP was launched in May 2000 to promote organic farming in India. India is member of United Nation and all member nations are directed to achieve the Millennium Development Goals (MDG) by 2020. For this purpose, the Government of India started a series of programmes and NPOP was one of them. In this direction a series of programmes were launched by the Indian government such as National Project on Organic Farming (NPOF), National Project on Management of Soil Health and Fertility (NPMSHF), Horticulture Mission for North East and Himalayan States (HMNEH), National Horticulture Mission (NHM), Network Project on Organic Farming under Indian Council of Agricultural Research (ICAR) and various other schemes under Agricultural and Processed Food Products Export Development Authority (APEDA). This was as the Ministry of Agriculture announced, 'Organic Farming Policy 2005'.

This chapter has two main sections: In section one the national organic policy and its provision are presented. The method of accreditation and certification are also included in this section. Detailed provisions of each component of organic farming is presented in this section. In section two the PGS-India scheme for certification is presented. The goals and methods of certification under PGS-India scheme, different stakeholders of this chain and cost effectiveness are discussed in this section.

NATIONAL POLICY FOR PROMOTING ORGANIC FARMING

The Ministry of Agriculture in India has not formulated any specific policy for organic farming. The first initiative was launched, the NPOP, by the Government of India in 2000. The NPOP policy draft has set up standards related to production, processing and handling for organic farming. All there organic farming activities require accreditation and inspection which were mainly facilitated by private certification agencies. The daft articulated rule of accreditation and certification as per international standards. The NPOP organic farming policy document has seven sections; definition and meaning of organic farming are in section one; scope and operational structure in section two; organic production, processing and handling standard in section three; accreditation and certification in section four and five; organic labelling requirement in section six; and list of permitted substances in section seven (NPOP, 2005).

The NPOP provides ethical and legal authority on accreditation and certification in India. The certifications are issued as per norms described in 'Reference Book' of Foreign Trade and Development Regulation Act (FTDR), 2001. The NPOP adopted these accreditation regulations on 25 May 2001. The National Accreditation Body (NAB) is the apex authority set up by the expert committee of the NPOP. The NPOP has notified the standards of accreditation and certification under FTDR Act which are controlled by APEDA. It works as a watchdog authority to meet the international norms for organic exports. The domestic certification matter deals under the Agriculture Produce Grading, Marking and Certification (APGMC) Act (Yadav, 2010). The organic products which are certified under APGMC by accreditation agencies are allowed to sell in domestic market. If accreditation agencies certified organic products and satisfied with quality as per NPOP standard, then only allowed to export (Morgera et al., 2012).

India has followed international organic regulations and provisions of EU regulation 2092/91 under Article 11(1) listed for Third-World countries. The certification standards are as per international standards and equivalent to Codex Guidelines, IFOAM and EU regulations. The Indian export of organic products have advantage to export in European Union Market. The Indian organic standards also meet the Switzerland organic regulations (Morgera et al., 2012).

Principles and Objectives

The NPOP policy is based on the four basic principles of organic farming and follows international standards for accreditation, cortication and inspection. The following are the objectives of the NPOP organic farming policy:

1. To set up standards for accreditation and certification of organic farming and products.
2. To develop mechanism for easy accessibility of accreditation and certification.
3. To formulate national standard for certification and accreditation organic products.
4. To encourage growth and development of farming through easy provision.

The policy document adopted provision of 'standards for standards' framework of IFOAM, which facilitates minimum standard for accreditation agencies. The four basic principles of organic agriculture are articulated in section 3 of NPOP policy document. These minimum provisions and principles are adopted in organic policy document.

The other provisions related to processing, handling and labelling are framed as per national and international standards. The four basic principles of organic agriculture are identical and other standards are as per basic standards of IFOAM, but it does not mean that they are the same. The NPOP policy document covered certification of land, method of production and product check for consumer satisfaction.

Certification of Land and Plant

Organic farming requires to make certification of land and plant with certification agency or with government certification body. Here plant means to the crops which grow must informed to certification agency. The certification and conversion of whole farm or livestock as organic require minimum 3 years, since first crop is harvested. The certification agency has authority to increase or decrease the time of conversion period of three years by inspection of soil status, but it should not be less than one year. In reducing conversion period, it must ensure and check that the producer has followed all standards. The certification body encourages farmers convert the whole farm and livestock rather than partially. If the whole farm is not converted, partial conversion is also allowed but with certain provision such as the producer must segregate land through natural barriers. Similarly, organic products, livestock products and other inputs should be separated essentially. Additionally, the organic converted land and livestock are not allowed to switch from organic to conventional (NPOP, 2005)

The NPOP document has a wide provision on management of organic land though compost, pest, disease and weed control norms, choice of crops, contamination control, soil and water conservation and collection of forest and wildlife products. It is mandatory for organic producers to use only certified organic seeds and inputs in crop cultivation. If organic inputs are not available in the market, they can prepare and use for cultivation. The organic policy does not allow use of pollen, genetically modified seeds and transgenic plants materials. The main objective of organic farming is to maintain fertility and biological conservation using crop rotation and biodegradable plants and animals. The quantity of organic inputs is determined by accreditation authority on the basis of type of land and agro-climatic conditions. The use of human excreta, synthetic nitrogen and nitrate are strictly prohibited (ibid., 2005).

The preventive techniques such as crop rotation, green manure and balanced use fertilizer are suggested to control diseases, weeds and pest. The physical methods including thermal sterilization of soil and products are used to control pest and disease. The use of synthetic and genetically modified organisms is strictly prohibited in organic cultivation. Some obligatory steps are required by the producers to minimize outside pollution of organic farms such as applying sustainable ways to enrich soil fertility and conserving water resource during production process. The clearing of land through burning of straw and clearing of primary forest is prohibited. Separate provisions are available in the policy document for land and livestock management, collection of forest and wild products. There is no specification of fixed time period for conversion of livestock into organic livestock, but partial conversion of livestock is only allowed when separated from conventional livestock (ibid., 2005).

Certification of Products

The certification of products has aspects of raw and processed product including livestock.[1] The products which are collected from hunting of wildlife are not considered as organic and excluded from livestock. The policy document followed general principles of labelling as per IFOAM Basic Standards under section II.B.7. The NPOP policy document classified organic products into four categories based on their organic composition.

The 'single-ingredient product with 100 per cent organic ingredients' are put under category-I. It means raw and processed products containing 100 per cent organic ingredients, excluding salt and water are labelled as 'produce of organic agriculture'. The 'multi-ingredients with 95 per cent organic ingredients' fall under category-II. It means same as category-I but percentage declines from 100 to 95 and certified as certified organic. The 'multi-ingredients with 70 per cent organic

[1] Livestock is defined as domestic animals that are used for food or for the production of food.

ingredients' fall under category-III. The meaning is same but the percentage declines from 95 to 70 and labelled as 'made with organic ingredients'. The 'multi-ingredients with less than 70 per cent organic ingredients' fall under category-IV and such products not given any label and are not considered as organic.

The logo for domestic organic product is 'Indian organic' and is used for all certified organic products identified as organically grown. This logo is only attached if the products are dually certified by accreditation and certification agencies. The remark of certification agency is voluntary, but using the logo is mandatory for all certified products and all accreditation agencies are allowed to use this logo (ibid., 2005).

Processing and Handling Regulation

The Indian standards for certification of handling and processing follow the general principle of organic products. The integrity of organic products should not be destroyed while handling, transporting and processing. The storage facility of organic products should be separate and not mixed with conventional and other prohibited products. In the processing of organic products ionising radiation is prohibited which is used for preventing disease and pest control. The processing of these products requires to preserve nutritional ingredients during physical, mechanical and biological processes. Hence the ingredients must be preserved in organic products when they are available for final consumption. The external addition of vitamins and minerals is strictly prohibited, though certification authority may permit to use non-organic minerals which are not genetically modified. The list of additives provided by certification agency which can be added during the processing of organic products, but it should be minimum and used when it is very essential. (ibid., 2005).

The following measures are used to control pests, disease and weeds in organic products. One is 'preventive measure'. In this measure the elimination and disruption of habitat and access of facilities are included. If the use of restricted pesticides is found during physical, biological and mechanical methods of processing then it should be not permitted to sell as organic. Furthermore, accreditation agency also verifies that approved additives ingredients and processing aids are used as per NPOP norms. For pest control in organic products disinfectants and persistent pesticides are not allowed. The accreditation and certification agencies allowed to set up chemical pest control testing lab.

In conclusion, Indian standards follow the international standards for storage, packaging and transportation norms for organic products. In packaging of organic products, the material used should be ecologically sound. The packaging of material should not affect the quality and integrity of the organic products. The substances harmful to human health are not allowed. The permitted condition for storage is specified as cooling, freezing, humidity and drying regulations (ibid., 2005).

Accreditation and Certification Agencies

The organic policy document has made criteria for accreditation and certification agencies and they must meet these criteria during inspection and certification. The National Centre for Organic Farming (NCOF) work as NAB and monitors these certification agencies. It approves the registration of these certification and accreditation agencies. The accreditation and certification licenses are open for individual, group, firm and cooperatives societies with condition to registered with national body within one year. Application for getting license must be submitted to APEDA and NAB. The processor of application, fee and documentary evidence of applicant must be submitted as per the prescribed format.

The accreditation license is permitted for following categories: (a) for organic producers, (b) organic product processing operators, (c) organic animal product producers and processing (d) wildlife animal organic products and (e) forest organic products. The accreditation agencies which are eligible for accreditation must sign accreditation contract with APEDA acting on the behalf of NAB. The accreditation number and certificate are not transferable. It is valid for three years from the date of issue and the renewal procedure is similar as getting initial accreditation. The renewal of license is based on subject to evaluation, review and annual surveillance report submitted by APEDA to NAB.

The evaluation committee also reports to NAB related to observed criteria of accreditation and certification programme (ibid., 2005).

The Indian standards for organic accreditation and certification are set at minimum standards. And these minimum standards of accreditation and certification are applied to both: method of production and processing chain. In processing the steps of handling of organic products are checked once at least annually. The accreditation agencies must follow the standard norms of inspection and certification of organic production and processing such as the same person will authorized for both certification and inspection. Along with this minimum requirement of accreditation and certification are monitoring, keeping records, obligations and reporting to higher authority. The inspection and certification fee charges vary for different operators and approved by NAB (ibid., 2005).

The certificate is issued by the concerned authority and in the Certificate of Registration all information such as personal details of operator, the products and validity is mentioned. The products are certified by assigned accredited agency as approved by NAB. The revision organic standards can be made by NPOP. The suspension and withdrawal of certification and accreditation license of agency are depending on violation and infringement of legal consequences without specifying code conducts which may lead one and other.

Organic Farming Policy, 2005

The organic farming policy was formulated in 2005 but its practice was continued by millions of Indian farmers since independent as traditional farming. The Indian farmers are using various types of natural manure for cultivation. Along with traditional habit of farmers, India has comparative advantage in organic farming due to lower cost of cultivation. The diverse agro-climatic conditions is another advantage for growing organically. The aim of organic policy is to promote economically viable, technically sound, socially acceptable and ecologically efficient method of production. The initial goal was to increase the certified area under organic farming. Other goals were to increase organic production, maintain soil fertility, preserve biodiversity, promote value addition, accelerate agri-business growth, strengthen rural economy and improve the standard of living of farmers and agriculture workers (GOI, 2005).

The organic farming policy has identified some thrust area: (a) to maintain soil fertility contents through biological cycle under organic farming. This process involves use of soil flora and fauna, micro-organisms, plants and animals. (b) To identify area and suitable crops for growing under organic farming. (c) To facilitate basic infrastructure for packaging and processing of organic products. (d) To establish model farms for getting organic seed for further use. (e) To make supply of organic seeds and inputs to producers available. (f) To apply traditional, biological and scientific methods for weed, disease and pest control. (g) To connect indigenous knowledge with scientific method to enrich organic farming. (h) To spread information regarding organic farming to other famers. (i) To facilitate organic market for selling organic input and output in domestic economy. (j) To raise income of organic producers by increasing productivity. (k) To generate employment opportunity for rural people. (l) To make easy inspection and certification process for domestic use. (m) To promote PGS system of certification. (n) To maintain diversity of ecosystem using natural inputs. (o) To improve condition of organic livestock and allow them to behave freely (ibid., 2005).

The policy document categorized existing 15 agro-climatic zones into three purity zones for organic cultivation. In the first category rain-fed and monocrops growing area, which have lower productivity using low chemical fertilizer are included. It is easy to covert these areas into organic certified area. States such as, Jharkhand, Rajasthan and Uttarakhand, and north-eastern regions falls under this category. The second category includes rain-fed areas having uncertain irrigation facility. In these rain-fed areas farmers are growing mono-crops that results loss of soil fertility which is main cause of lower productivity. States such as Himachal Pradesh, Chhattisgarh, Gujarat, Madhya Pradesh, Orrisa, Jammu and Kashmir and some parts of Maharashtra and Karnataka fall in this category. In the third category areas using massive chemical fertilizer under multiple cropping are included. The conversion of these areas into organic farm may lead to loss in productivity. The balance use of manure, biomass, inorganic fertilizers and pest management helps to achieve agriculture sustainability (ibid., 2005).

In the policy document it is accepted that India has very negligible share of organic market in production and trade; but it is seen as an emerging sector, showing presence in the organic world market in products such as cotton, spices, tea, coffee, oilseeds, cereals, pulses, fruits and vegetables. The major crops growing under organic cultivation are vegetables and fruits such as mango, orange, guava, papaya, passion, cashew, pineapple, walnut and mausambi. The document also talks about generation of rural income and employment via adopting organic method of cultivation

The method of organic cultivation is sustainable and is based on locally available manure and farm inputs. The policy document stated that farmers are encouraged to prepare manure, vermicompost, blue green algae and azolla at own farm. When it is preparing locally, the cost of transportation and packaging will be zero, which leads to reduced manure price. The document emphasizes on local knowledge on ecology and local method of compost and seed preparation combine to make farming sustainable. The increasing demand of organic products will develop organic market and help to adopt organic farming by more farmers. The price of organic products is higher than the conventional, which attracts more farmers to adopt organic farming to achieve one of the policy goals (ibid., 2005).

There is opportunity to engage in accreditation, certification and marketing of organic products. The opportunity of employment at home can play an important role in development of domestic organic market. The local bazar is indicative taken by the local individuals for selling and buying organic products at village. Organic fair may be organized once in week to attract consumers towards organic products. These organic small markets can also link with organic Agri export zones. The cost effective PGS system of certification can help small and marginal farmers to adopt organic farming. The policy document initiated this system of certification through local group, regional council and national council.

CERTIFICATION PROCESS AND PGS-INDIA PROGRAM

The policy makers received the feedback from researchers and farmers on the method and cost issues of accreditation and certification. The found that third party certification and inspection method is not appropriate for a country like India where majority of farmers are small, marginal and medium. Then inclusive, affordable and cost-effective certification PGS was developed. The acceptance of third-party certification for producers, distrusters and organic products are popular and charging high cost for certification. This is based on audit trail for steps of cultivation, seeds, manure and final products. The paperwork is more complicated in this system and difficult for Indian farmers, where most of the farmers are illiterate, small and marginal. A three days' workshop on PGS for India was conducted on 23–25 September 2006 in Goa by FAO.

A pilot project was conducted, 'A Participatory Organic Guarantee System for India' in 2006. This project was conducted in 14 councils of 10 states for a period of six month. Three members Mathew John, Claude Alvares and Joy Daniel team were constituted for documentation process required and implementation of this system of certification. Furthermore, a council of 14 members closely works with above committee to finalize the standards. Now the PGS system became most popular systems among the Indian farmers. An essential expense and simple paperwork are required under PGS system. The certification and accreditation are available for organic farmland, method of cultivation, formation of manure and so on. The aim of this cost-effective system to promote organic farming movement in India (FAO and Samanvaya, 2006).

The individual efforts also made to the cost of accreditation and different methods were suggested by the NGOs and societies. The collective efforts of all individuals, civil societies and NGO's developed PGS system. Here the term 'participatory' embodies the active participation of producer, consumers and other stakeholder of organic farming. The NPOP has defined PGS as a viable alternative system of certification, in which a group of farmers monitor each other and collective registered with regional and national council of India. The IFOAM (2008) also accepted the PGS system of certification and defined as 'PGS are locally focused quality assurance systems. They certify producers based on active participation of stakeholders and are built on a foundation of trust, social networks and knowledge exchange.'

The PGS-India follows the IFOAM norms and accepted the ecological approach under which use of genetic modified seeds, chemical fertilizers and pesticide are prohibited. This system of certification concentrates small and marginal farmers, those who want to adopt organic farming but unable to pay huge cost of certification under third party certification. It is based on community and cooperation approach and the goal is the same as third party system to provide guarantee for consumers on quality of organic products. The method of production and quality assurance are two main goals of the PGS from the consumer point of view. This method is more realistic and collective participation of farmers empowers them. This collective approach helps farmers, enhance knowledge on manure preparation, soil preparation, water and seeds management, management of livestock, pest and disease control. The active participation of producers results less record-keeping and paperwork. It is based on integrity approach with its foundation was trust and trust developed in the group with openness equality and transparency in likeminded environment that reduces administrative hassle (FAO and Samanvaya, 2006).

The PGS-India report has mentioned six basic elements of this system: (a) *Participatory,* it is based on the idea of collectiveness and strong participation by the organic producers and consumers. The principle and rule of organic farming are applicable to all stakeholder's collectivity. (b) *Share vision*, it is based on fundamental value of sharing information among the producers and consumers in guiding basic principle of organic farming. (c) *Transparency*, it is based on the guiding principle of every information is available for all stakeholders. (d) *Trust*, it is based on faith and that consumers can trust on organic product through certification (e) *Horizontality*, it is based on power sharing means all stake holders having same ability and responsibility to assure product quality and (f) *National networking,* it is based on interlinkage among the stakeholders from farmer to local group and local group to regional council and regional council to national council (PGS-India, 2015).

Role and Responsibility of Stakeholders

There are different stakeholders which are playing their role and responsibility in organic farming movement. Thus, the healthy relations among the different stakeholders is a crucial task for regional and national coordination committee. The producer or farm family is the key stakeholder in the whole organic movement. So, the understanding of organic farming standards to the farmers and its family members is essential for active participation in organic production. The family members of producers will perform all the agriculture related activity so their training must be organized by regional council at local level. The members are required to attend and share information with each other. The consumers are also allowed to visit at farm and suggest what will need to more (FAO and Samanvaya, 2006).

The figure 14.1 shows that the responsibility of different stakeholders different and the size of box presents the responsibility of the organic movement. Because the proportion of responsibility depends on the size of box so the local group has the highest followed for the farm family then regional group and least responsibility is borne by the national authority under the PGS certification system. In the PGS system local group and farmers are the key stakeholders, farmers can

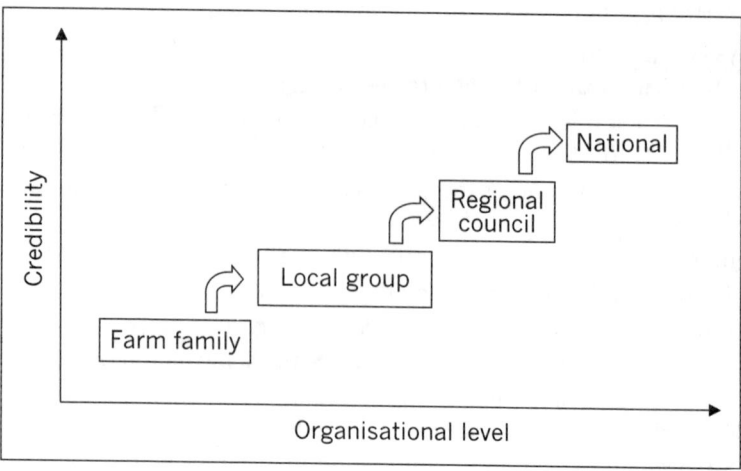

Figure 14.1 *Key Stakeholders*

make their separate group by registration with regional council. The NGOs, government body and local farmers group those able to facilitate accreditation and certification can get permission from regional council to set up PGS. The functioning of local group is easier than individual certification and is cost effective. It is not a compulsion that all member of local group should be certified as organic farmers, few may not meet the requirement; but if they willing to join, they can join. The local group works as a service provider and functions as mediator between farmer and regional council. It has power to decide the year of certification, act as default and coordinate with regional council for inspection and appraisal of farm (FAO and Samanvaya, 2006).

The regional council is a state level authority, but the local group has the option to form their own regional council separately. The role and function of regional council are to conduct training workshop programmes on organic farming with the coordination of national council. This is chain-based system, where all the local groups are registered with regional council and all regional council with the national council. The certification IDs are generated for individual farmers after the approval of all kind of paperwork by local group.

The national coordination committee represents, the Ministry of Commerce, Ministry of Agriculture, APEDA, NGOs, consumers groups and qualified members of regional councils having 1–2 years' experience. The main function of this committee is to coordinate with regional councils, approve new regional councils and conduct training workshop. Its basic responsibility to regulate all regional councils and local groups as per basic standards of standards based on four principles of organic farming. It maintains database such as pool of IDs numbers of regional, local and individual famers. It also links with the APEDA the exporting authority of organic products.

PRESENT STATUS OF PGS-INDIA

The council for PGS Organic India was set up in 2006 after a common consensus. The function of this council was to work for the promotion of organic farming in India. To meet the domestic demand of organic product was the one of the objectives of council. The Government of

India launched PGS India scheme in 2011, which was initiated by different organization which are working for promotion of organic farming. The Participatory, Guarantee System Organic Council (PGSOC) at Goa was recognized as formal PGS registered society. It has deigned two logos for identification of organic products one is 'PGS-India Organic' and another is 'PGS-India Green'. The 'PGS-India Organic' means the products are produced under certified organic farmers and they have complied their conversion period. On the other hand, 'PGS-India Green' means products are produced those farmers are in conversion period.

In India more than 20 third party certification agencies were working in the states such as, Andhra Pradesh, Haryana, Karnataka, Maharashtra, Rajasthan, Uttarakhand and Kerala, till 2015. India has set up seven centres one national and six regional centres for training, promotions and guidance of organic farmers. The national centre for organic farming in Ghaziabad, Utter Pradesh, serves for states of Delhi, Rajasthan, Uttarakhand and Uttar Pradesh. The Regional Organic Centre of Bangalore in Karnataka covers states of Tamil Nadu Karnataka, Pondicherry and Lakshadweep. The Regional Organic Centre of Bhubaneshwar in Odisha serves states of Odisha, Bihar, West Bengal, Andaman and Nicobar. The Regional Organic Centre of Hissar in Haryana covers states of Panjab, Haryana, Jammu & Kashmir and Himachal Pradesh. The Regional Organic Centre of Imphal in Manipur serves states of Assam, Arunachal Pradesh, Mizoram, Meghalaya, Manipur, Sikkim, Tripura and Nagaland. The Regional Organic Centre of Jabalpur in Madhya Pradesh covers states, Chhattisgarh, Jharkhand and Madhya Pradesh. The Regional Organic Centre of Nagpur in Maharashtra serves states Andhra Pradesh, Gujarat, Maharashtra, Goa, Daman & Diu, and Dadra Nagar Haveli (NPOF, 2005).

Community-Based Scheme for Organic Farming

The government of India has announced some specific schemes focusing on specific areas for *paramparagat krishi* (traditional farming) and sections of society such as women for the promotion of organic farming in India. These two schemes are specific in nature and started as pilot project in chosen states with collaboration of state, society and self-help groups. These two schemes are Mahika Kisan Sashaktikaran Pariyojana (MKSP) in 2010 and Paramparagat Krishi Vikash Yojana (PKVY) in 2017.

Mahila Kisan Sashaktikaran Pariyojana (MKSP)

Since ancient time women's participation in agriculture activity has been crucial and all the other activity except ploughing of field is done by women in India. Women are cited as backbone of agricultural labour force across the world but her hard work in most of the cases are unpaid. Women have played very important role in agriculture extinction activities such as Farm activities, which includes transplanting and harvesting; past harvest activities which cover, threshing, drying and parboiling and livestock management which includes shed cleaning, fodder collection and mulching. In all these activities the role of women is very important (ICAR, 2014). The above contribution of women in agriculture has forced government to announce a specific scheme for women famers. As a result, Government of India announced MKSP as a subcomponent of the National Rural Livelihood Mission (NRLM) with financial assistance of Rs 100 core in 2010–2011 annual budget. It is implemented in 15 blocks of 9 districts of Bihar with 1,400 villages in 2012, with total cost of 96.70 crore and success of MKSP is getting popularity across the states in India.

Objectives: The primary objective of the MKSP is to empower women farmers by making systematic investments to enhance their participation and productivity under sustainable agriculture. Some specific objective of MKSP are: (a) To enhance agriculture productivity through women participation. (b) To create sustainable agriculture livelihood opportunities for women. (c) To improve women farmers skills and capability to support farm and non-farm activities. (d) To ensure food and nutritional security. (e) To provide complete access inputs and financial services for women. (f) To enhance managerial capabilities of women farmers for management of biodiversity. (g) To improve accessibility of institutional services and schemes for women farmer.

Approach and Strategy: The approach of MKSP is women farmer centric and provides direct and indirect support to achieve sustainable agriculture production. It will help women to learn and adopt new sustainable farming technology. The skill and capability development of women farmers are enhanced through learning by doing process. The following strategies are adopted to implement MKSP in true spirit. (a) Use locally adopted, resource conserving, local knowledge-centric and environmentally friendly farming techniques. (b) Coordinate with women self-help groups, NGOs and farmer groups, agriculture university and Krishi Vikash Kendra for sustainable agriculture practices. (c) To spread the benefits of sustainable agriculture methods among the women through demonstration. (d) Enhance women skills through capacity building, formal and vocation training courses. (e) It will help poor women and all social groups women. (f) Identify the target group, special women headed household and provide financial and technical support through participatory approach.

Paramparagat Krishi Vikash Yojana (PKVY)

The Paramparagat Krishi Vikash Yojana (PKVY) was launched in 2017, as extended component of Soil Heath Management (SHM) scheme already working under the centrally sponsored National Mission on Sustainable Agriculture (NMSA). The aim of PKVY is to promote organic farming to improve soil heath as result of organic farming. The PKVY also promotes PGS--India programme for organic certification through mutual trust. The financial support for scheme PKVY is 60:40 ratio between central and state governments. In the case of north-eastern and Himalayan states the financial ratio will be 90:10 per cent and for union territories it will be 100 per cent financial assistance by the central government (PKVY, 2017).

Objective: The objective of PKVY is to produce chemical-free agricultural products through low-cost technologies and organic farming. The key point of PKVY for promoting organic farming are: (a) To promote organic farming in rural area, (b) To disseminate latest technology in organic farming, (c) To explore the agriculture research and expert knowledge to increase production under organic farming,

(d) To organize at least one demonstration on organic farming in a village at a cluster.

Approach: The PKVY implemented in a three-year time bound framework under the PGS-India scheme. It is a scheme for community-based organic farming certification system of 36 months conversion period from conventional to organic. PKVY promotes organic farming through a cluster approach under to access PGS-certification. The PGS certification system facilitates farmers to register their farms at nearby village PGS and to certify their produce, level their product domestically.

Main feature of the PKVY Scheme: (a) The cluster size should be 20 hectares and contiguous chosen for organic farming. (b) The maximum Rs 10 lakhs financial assistance available for a cluster size 20 ha and Rs 4.95 lakhs for mobilization and PGS certification with a subsidy ceiling of per hectare per farmer. (c) In each cluster, a minimum 65 per cent farmers should be small and marginal. (d) The budget allocation under PKVY need to allocate at least 30 per cent budget to women beneficiaries/farmers.

Key Components of PKVY: The PKVY has two main components: Model Organic Cluster Demonstration (MOCD) and Model Organic Farm (MOF). The MOCD aims to create awareness among the rural youth, farmers, consumers and traders on latest technologies of organic farming. These demonstrations are conducted at organic farm field cluster through zonal councils or National Center for Organic Farming (NCOF). The major two activities are conducted under the MOCD, one is adoption of PGS certification and another is manure management and biological nitrogen harvesting. In demonstration farmers are guide for PGS certification process and quality control of organic products under first activity. The conversion of conventional land into organic land, integrated manure management, custom hiring centres and packing, labelling and branding of produce are done under the second activity. The aim of MOF is to present demonstration conversion of conventional land into organic practices in one-hectare area to other farmers. This demonstration spared the information on latest technologies and organic input production process through exposer visits of famers (PKVY, 2017).

CONCLUSION

India has formulated a national policy for organic farming under the ministry of commerce and made some provisions for organic producers. These provisions are related to certification, inspection and use of seed, manure and organic method of pest control. The national centre for organic farming is providing technical support to farmers with the help of six regional centres. These centres organize training programmes for farmers to guide manure preparation, irrigation and seed-related problems. It is found that reducing the high cost of certification is a major task and for this PGS-India was introduced across India. The PKVY scheme may help to implement PGS in the true spirit.

REFERENCES

DARE/ICAR (2013). *Women and agriculture.* New Delhi: Indian Council for Agriculture Research, Annual Report, 2013–2014.

FAO and Samanvaya (2006). Participatory Guarantee System (PGS) for India, Proceedings of Workshop held in Goa, 23-25 Published in September 2006.

MKSP (2010). *Mahila Kisan Sashktikaran Pariyojana* (policy draft). New Delhi: Government of India.

GOI (2005). *Organic Farming Policy 2005*, policy document, Ministry of Agriculture, Department of Agriculture & Cooperation.

Morgera, E., Caro, B., & Durán, M. (2012). *Organic agriculture and the law.* Rome: Food and Agriculture Organization of the United Nations.

NPOP. (2005). *National program for organic farming* (6th ed). New Delhi: Department of Commerce, Ministry of Commerce & Industries, Government of India.

PGS-India. (2015). *Participatory guarantee system for India: operational manual for domestic organic certification.* New Delhi: Ministry of Agriculture Department of Agriculture and Co-operation National Centre of Organic Farming, Government of India.

PKVY (2017). *Paramparagat Krishi Vikas Yojana—manual for district level functionaries*, 2017. New Delhi: Department of Agriculture Cooperation and Farmers Welfare, Government of India.

Yadav, A. K. (2010). *Organic agriculture, concept, scenario, principals and practices.* Uttar Pradesh: Director National Centre of Organic Farming, Ghaziabad National Centre of Organic Farming Department of Agriculture and Cooperation, Ministry of Agriculture, Government of India.

State Policy and Regulation

INTRODUCTION

India is a union of states and the Constitution of India provides division of power between the centre and states. The division is mentioned in its Seventh Schedule, which the Parliament has the power to legislate on. The legislative section is classified in to three lists, union, state and concurrent list. Each list has a certain number of subjects for which they can make laws and rules for development. The central list has the largest number of items 100 (the last item numbered 97) followed by state having items 61 (the last item numbered 59) and concurrent list having 52 items (the last numbered 47). Agriculture is a state subject so the state has freedom to design separate policy and provision for the development and growth of agricultural sector. As per the variation of agro-climatic zone, some states have more potential for organic farming, such as north-eastern states and hill states such as Himachal Pradesh, Jammu and Kashmir, Uttarakhand and Madhya Pradesh. Having climatic benefit to adopt organic farming, in these states the productivity difference between conventional and organic farming is minimum.

Sikkim is the first state which drafted the organic farming policy in 2003. After that followed Mizoram (Government of Mizoram, 2004), Karnataka (Commissionerate of Agriculture, Government of Karnataka, 2004), Kerala (Department of Agriculture, Government of Kerala, 2008), Andhra Pradesh (Department of Agriculture, Government of Andhra Pradesh, 2008) and other states drafted policy 2010 onwards. The objectives of all states organic farming policy are near about similar and followed the four basic principles of organic farming as per objectives of national organic farming policy.

The certification and inspection provision cost are vary from state to state but the norms are similar. Some states provide compaction of losses due to adoption of organic farming to formers in conversion period of three years. These losses are due to initial falls in productivity due to abandonment of chemical fertilizers and use of organic compost. In this chapter the state-wise progress on formation of organic policy and the detailed policy provisions for the state of Gujarat, which has recently designated its organic farming policy document in 2015 are provided. The zero budget natural farming (ZBNF) is a method of farming under which cost of growing and harvesting is zero. It is getting popularity in states such as Maharashtra, Karnataka, Andhra Pradesh, Himachal Pradesh, Uttarakhand, Chhattisgarh and Kerala, where millions of farmers are adopting this method.

STATE-WISE ORGANIC FARMING POLICY

Indian constitutional provisions mandate agriculture as a state subject; it means that the states have decision-making power for collection of tax from agriculture and implementation of any land-related rues and regulation. India has also followed other advanced countries that are leading organic farming movement after 2000. The central government can formulate agricultural policy for the nation as a whole, but if states are not interested to adopt it, then national policy fails to achieve their goals, because agriculture is state subject. Same is the case of organic farming policy. India formulated the draft policy in 2005 but most of the states did not implement it. Some states that are leading in adopting organic farming formulated their own farming policy, namely, Andhra Pradesh, Karnataka, Madhya Pradesh, Maharashtra, Tamil Nadu, Himachal Pradesh, Sikkim, Kerala and Gujarat.

An early step in policy formulation was taken by Sikkim in 2003, followed by Karnataka in 2004, although the central government policy for organic farming was not deigned yet. It was designed within the agriculture policy in 2005. After that other states began to think in this direction. Then all other states set up their own expert groups to make drafts for organic farming. Table 15.1 presents 10 states of India that formulated their own state policy for organic farming for the promotion of organic farming in their own state.

Table 15.1 Comparisons of States Policy Impacts on Organic Cultivation Area (ha)

States	Year	Objectives	2005–2006	2013–2014	CAGR	CV
Andhra Pradesh	2008	10	1,661.42	12,325	13.63	0.75
Himachal Pradesh	2011	05	3,647.41	4,686.05	77.78	0.99
Gujarat	2015	13	1,627.06	46,863.9	30.87	0.99
Kerala	2008	11	15,474.47	15,020.2	−1.63	0.27
Karnataka	2004	10	4,117.17	30,716.2	25.59	0.72
Madhya Pradesh	2010	02	16,581.37	232,887	43.85	0.65
Mizoram	2004	05	300.4	0	5.17	0.65
Nagaland	2013	11	718.76	373.13	7.96	1.06
Sikkim	2003	08	177.64	60,843.5	105.62	1.00
Chhattisgarh	Draft	–	293.16	4,113.25	99.25	1.89
Maharashtra	Draft	–	18,786.69	85,536.7	95.37	0.65
Goa	Draft	–	5,555.07	12,853.9	15.76	0.35
Tamil Nadu	Draft	–	5,423.63	3,640.07	19.15	0.38

(Table 15.1 Continued)

(Table 15.1 Continued)

States	Year	Objectives	2005–2006	2013–2014	CAGR	CV
Uttaranchal	Draft	–	5,915.85	24,739.5	36.81	0.57
Rajasthan	–	–	22,104.91	66,020.4	41.84	0.52
Orissa	2018	03	26,387.86	49,813.5	–6.07	0.59
Uttar Pradesh	–	–	3,033.98	44,670.1	65.57	0.67
India	2005[a]	03	173,682.54	723,039	35.06	0.47

Source: States Policy Documents and NCOF Annual Reports.
[a] Ministry of Agriculture, Government of India policy provision, 2005.

The goals of all states' organic farming policy are the same, having their own provision for certification and inspection and other standards, but all have followed national standards. Table 15.1 presents the year of organic farming policy and compound annual growth rate (CAGR)[1] of certified organic land during 2005–2006 to 2013–2014. The value of CV shows that if it is more than 1, it has more variability in growth and if less than 0.5, then that means less variability in data. In the table is found that most of state recorded less than 0.5 value of CV showing consistency in growth for 2005–2006 to 2013–2014.

ZONE-WISE ORGANIC FARMING POLICY IN INDIA

India is divided into six zones based on climatic, geographical and cultural features namely, north zone, east zone, south zone, west zone, central zone and north-east zone. These six zones comprise 29 states and 7 union territories and each zone has certain number of states and union territories. For the state-wise analysis of organic farming policy, one state is chosen from each zone: Himachal Pradesh from north zone, Orissa from east zone, Madhya Pradesh from central zone, Gujarat from western zone, Karnataka from south zone and Sikkim from north-east zone across the India.

NORTH ZONE (HIMACHAL PRADESH)

North zone comprises six states of India, namely Jammu and Kashmir, Himachal Pradesh,

Haryana, Punjab, Uttarakhand and Uttar Pradesh. For the study of state organic farming policy Himachal Pradesh is chosen because other states have not formulated state organic farming policy yet. Some of them formulated policy drafts and other states are in the process of formulating state organic farming policy. Himachal Pradesh is a hilly state and agriculture is a lifeline of the state economy where nearly 70 per cent of the population is dependent on this sector. Presently,

---

[1] Annual Compound Growth Rate (ACGR) calculated using semi log model, for reducing fluctuation first converted value into long vale form and then applied formula for calculation of ACGR = EXP(β) − 1.

Table 15.2 *State Organic Farming Policy from Each Zone*

Zone	State	Year	Objectives	Provisions
North	Himachal Pradesh	2011	05	To achieve four principles
East	Orissa	2018	03	Climate resilient, reduce risk and enhance farm income
West	Gujarat	2017	13	Area specific crop cultivation
South	Karnataka	2004	10	Integrated approach
Central	Madhya Pradesh	2010	02	Health food for all
North-east	Sikkim	2003	08	Sustainable farming practice

Source: States policy documents.

agriculture sector in Himachal Pradesh is passing through a transitional phase and farmers are looking for alternative farming methods like organic farming. Himachal Pradesh is leading in production of organic vegetables and fruits, and farmers' income transforms drastically by adopting organic farming.

Himachal Pradesh formulated its own organic farming policy in 2011 having one main objective and five sub-objectives. The preamble of State organic farming policy would strive for adoption of organic farming practices and will bring about policy changes in order to provide sustainable livelihoods to the farming community through organic farming. This policy document adopted all four basic principles: the principle of heath, principle of ecology, principle of fairness and principle of care for achieving goal of organic farming in the state.

Objectives: The goal of Himachal Pradesh organic farming policy is 'to see that the dream of organic vision come true, it is imperative to undertake all necessary initiatives, including':

1. Giving recognition and encouraging growth of organic sector and developing agriculture sector in the state.
2. Creating enabling environment for the growth and development of organic farming by facilitating input and output organic market facility in the state.

3. Making policy provision to strengthen crop-livestock linkages for making it an organic compost-rich state.
4. Making easy certification process and facility to each farmer.
5. Creating investment environment for organic agribusiness and organic villages/valleys based organic agro-tourism.

Vision: The long-term vision of Himachal Pradesh organic farming policy is 'Organic Himachal Pradesh' whereby practicing it, economic and ecological benefit would be received by all farmers. Farmers and people of the state enjoy pesticide-free safe food, vegetables, fibre, fruits, milk and water. Many youths have become elf-employed through organic agribusiness and organic villages based agro-tourism. The state will contribute to reduce carbon footprints, environmental health and agro-biodiversity conservation.

Strategies: The strategies for development of organic farming in the state are mentioned in policy document, such as (a) reforming agriculture and allied sector policies to bring these in line with organic vision, (b) state will formulate organic policies and strategies for farming concerns/needs, potential niches, (c) government support to organic sector, (d) support to the farmers by providing financial help, (e) the incentive input schemes for farmers and (f) set up Himachal Organic Farmers Forum.

Implementation: For speedy and effective implementation of the state organic farming policy, the state government will ensure an appropriate institutional mechanism and facility. Both short-term and long-term actions required to achieve policy goals would be suggested by a full-time team of professionals having requisite expertise and experience. Universities would be encouraged to research on organic farming and link with the policy and practice.

EAST ZONE (ODISHA)

East zone comprises four states of India, namely, Bihar, Odisha, Jharkhand and West Bengal. For the study of state organic farming policy Odisha is chosen because other states have not formulated state organic farming policy yet. Government of Odisha announced

Odisha organic farming policy in 2018 and intends to build upon the natural agro-ecological and bio-diversity advantage to harness growth potential of sustainable and climate resilient organic agriculture. In the next five years, the state intends to identify, map and bring 2.0 lakh ha of agriculture area including horticulture and forest into organic farming.

Objectives: The organic agriculture policy of Odisha intends to achieve the following:

1. To provide healthier, diverse, nutritive and chemical-free food for local consumption in rural and urban areas.
2. To promote healthier soils and viable farms with agro-ecological approaches to farming substantially reducing the input costs and enhancing environmental services.
3. To promote and harness the growth potential of organic markets, both internal and external.

Focus Area: Four strategic focus areas targeted for the purpose are as follows:

Stream 1: Default organic areas of the state. The state with low penetration of
chemical-based technologies will be delineated for strategic interventions under Stream 1.
Stream 2: Areas where use of chemical fertilizers, herbicides and pesticides is moderate and INM practices are being promoted.
Stream 3: Areas where organic certification is going on targeting organic export markets and niche markets.
Stream 4: Areas in which using high amount of pesticides and chemical fertilizers.

Promotion for Organic Farming: The promotion and advertisement of organic farming is significant aspect of the Odisha state organic farming policy. it facilitate, Knowledge Dissemination, Soil Health Management, Availability of quality seeds, to avail agricultural credit, Specific focus will be on avoiding mono crops and promotion of diverse crop systems, The mechanization requirements for organic farmers will be assessed locally, facility for post-harvest management of organic produce, Livestock and Fisheries Development in Organic landscapes, Harnessing growth potential of default organic areas including forest and non-timber products. Government of Odisha will identify and declare suitable forest areas as organic zones and develop time-bound plan for replacing chemical fertilizers and synthetic pesticides with environmental-friendly approaches.

Market Development for Organic Produce: The government will ensure to consumers faith on quality assurance through certification and labelling. The assurance of quality and organic produce is crucial for market development of organic products which are made through certification and labelling. The state organic farming policy will promote both PGS and third-party certification type certification for domestic and export purpose. Another additional facility for quality test is to develop regional pesticide residue testing laboratories for the purpose of promotion of organic products. Try to reduce the cost of certification and inspection that will help famers to adopt organic farming. For marketing of organic products, appropriate platforms including investment in supply chains, certification, packaging, e-markets and so on would be developed. Necessary support will be provided for export promotion of organic products.

Research and Extension: For research and development of organic farming, it evokes to facilitate research and experiment facility in universities and institutions. It also advocates to organize training programmes for organic farmers with cooperation of regional center and relevant departments.

Odisha Organic Mission: Odisha Organic Mission (OOM) will be established by the state government as an institutional mechanism to coordinate various efforts that complement promotion of organic farming in the state. The mission will also be responsible for developing necessary programmes and operational mechanisms for implementing various policy and programmes under organic farming. The institutional mechanisms accomplish the following functions: (a) Delineate the three types of niche areas (Streams 1, 2 and 3) with high potential for organic farming. (b) Bring out necessary guidelines for declaring the organic zones at the village, GP, block and district level. (c) Coordinate the efforts in promotion and strengthening of organic farmers' collectives that include empanelling and supporting the resource organisations. (d) Approval of annual budgetary provisions and programmes.

WEST ZONE (GUJARAT)

West zone comprises seven states and union territories of India, namely, Rajasthan, Maharashtra, Goa, Gujarat, Dadra and Nagar Haveli, Daman and Diu. For the study of state organic farming policy Gujarat is chosen because other states have not formulated state organic farming policy yet. The government of Gujarat has recognized that there is a lack of organized market support system of organic inputs and outputs, professionally trained human resources, produce quality assurance, demonstration and training facilities, which restrict the growth of organic sector. For the documentation and data base, limited efforts have been made. It is widely observed that there is need of 'hand holding service' during the conversion period, initially for three years, including technical, market support, financial support, social and emotional support. At the initial stage organic farming policy may be helpful in increasing organic crops production, horticulture crops and animal husbandry.

Aims and Objectives: The Gujarat organic policy document (Department of Agriculture, Government of Gujarat, 2015) presented its own view to achieve the state agriculture suitability by adopting organic farming in true sprit through trusting both buyers and producers of organic products. Its aims are pointed out as economically viable, eco-friendly and technically sound to achieve agriculture sustainability for producers as well as ecosystem. Policy document presented long-term vision for proper value chain processes from which produces and consumers will benefited. The main gaol of policy document is to increase certified by ten times in the next few years. Following are some objective which are borrowed from the Gujarat state policy document 2015:

1. Maintenance of soil fertility under organic farming through biological process of cultivation using compost and micro-organisms.

2. Specified organic crop for classified area.
3. Facilitate organic inputs facilities and encourage to use it.
4. Set up ideal model farm for organic seeds production.
5. Assurance of quality check for both inputs and output.
6. Assured use of natural method to diseases and pest control.
7. Use biological method for weed control.
8. Promote traditional and indigenous knowledge.
9. Make farmers aware about organic farming.
10. Facilitate organic products market facility to farmers.
11. Improve farmers income by selling at higher price.
12. Promote cost effective group certification.
13. Facilitate regulatory norms for organic outputs and inputs

Principles and Definition: The organic farming Gujarat policy document (Department of Agriculture, Government of Gujarat, 2015) has quoted some international organization which defined organic farming like Codex Alimentarius Commission, whose definition reads, 'It is a holistic cultivation method which manage ecosystem via biological activity.' This definition puts emphasis on off-farm management activity based on regional and climatic conditions. The process of cultivation follows the biological and agronomic method using manure.

Another definition the policy document quoted is given by IFOAM as 'It is method of cultivation that sustain ecosystem, health of soil and mankind. It relies on biodiversity, ecological processors, and crop cycle as per local condition, instead of using unwanted inputs. It combines science, innovation and traditional knowledge for betterment of environment and quality output.'

In the same line the policy document also quoted the definition of NPOP as 'a method of cultivation that manage ecosystem and production sustainability simultaneously without using genetically modified seed, chemical fertilisers and pesticides'. Finally, the document also pointed out definition of Sajiv Kheti as 'it is method of farming that preserve natural resources along with producing enough that feed the need of local, present and future generation.'

Thrust Areas: The organic policy document of Gujarat (Department of Agriculture, Government of Gujarat, 2015) has pointed out some thrust areas which must be focused to attain better results. These areas are farm ecology and landscape management by the producers, supervision, organic seeds, organic input management along with bio-fertilizer, diseases and pest control through biological methods, plant health management and neutral method of weeds controls. The document has also insisted on some other thrust areas that would help in the promotion of organic farming. These are related to facilitating formal channels for credit and financial accessibility that help in pro-duction of organic inputs, marketing, research and development, along with generation of employment. The accreditation and certification require some minimum amount of funds that help to adopt organic farming in true spirit. The human resource development is another area that must be achieved through education and proper training of farmers and all stakeholders of organic agriculture.

Implementation of Policy: The organic policy document of Gujarat state (Department of Agriculture, Government of Gujarat, 2015) has made some recommendation for better result and policy implication. One the agro-climatic condition and type of soil are decision-makers to specific crop for organic cultivation. The division of agro-climatic was adopted as per state classification in different regions. Eastern Gujarat having districts such as Chota Udaipur, Dahod, Dang, Narvada, Panchmahals, Sabarkantha, Surat and Tapi have the most potential for organic cultivation. Then the policy document further highlighted the division of Gujarat state into eight agro-climatic zones having opportunity for growing different types of organic plants and crops. The soil type and climatic condition such as rainfall are main determinants of agro-climatic zone and crop specification. The policy document classified six categories on the basis organic farming oppor-tunity. First is category-I organic farming conducting by default that can be easily converted, second is category-II specified as niche area, then category-III as using lower external input in cultivation, category-IV is the most productive area, then category-V as animal products and finally category-VI institutional experiment field.

Policy Initiatives: The organic policy document of Gujarat state (Department of Agriculture, Government of Gujarat, 2015) has pointed out some initiatives that can help to increase the growth of organic farming in the state. These initiatives include providing some organic input incentives in the conversion period for three years. In this period the productivity may fall so financial assistance is essential to farmer to stay with organic farming or else he/she may quit organic farming. At the same time if the price of organic products remains stagnant, then government must be compensate the loss incurred by farmers. Another initiative also relates to purchase of organic seed, manure, organic compost, etc. These are some supply-side obstacles that may be present in the growth of organic farming in the state. Policy document also suggested community-based organic input production units which can supply organic compost production and bio-energy. To increase the supply of organic inputs, there is a need to set up some organic production units along with proving facility for certification, documentation and easy accreditation.

The organic policy document of Gujarat (Department of Agriculture, Government of Gujarat, 2015) also proposed to set up some model organic farming and resources centres in existing agricultural universities in the state. Those universities that have their own research and academic experience can develop organic research centres for the farmers. For sustainable agriculture organic farming may be the only solution. It can be made more productive through education, training and learning by doing. Policy document has also proposed that agricultural universities of state can offer few short-term courses that will help to spread information in the society. The Krishi Vikas Kendra (KVK) can be used as training centres for farmers, input producers and certification processors. The policy document also suggested that a part of certification and accreditation cost state may be subsidized. It should be from 25 per cent to 75 per cent of total cost of certification and inspection and promote PGS system of certification.

SOUTH ZONE (KARNATAKA)

South zone comprises four major states of India namely Andhra Pradesh, Karnataka, Kerala and Tamil Nadu as well as he union territory of Puducherry. For the study of state organic farming policy Karnataka is chosen because it is the pioneering state on organic farming in south zone. Karnataka is bestowed with varied climatic and soil types spread across 10 agro-climatic zones. The state of Karnataka has geographical opportunities, namely, Western Ghats, coastal plains and plateau allowing it to cultivate a variety of organic crops. The annual average rainfall of Karnataka is 1,130 mm. The state has moderate temperature along with other climatic condition which provide conditions to cultivate various crops in all seasons. Animal husbandry and horticulture provide support to grow many other crops organically. Karnataka is one of the pioneering states which announced its organic farming policy in 2004 after Sikkim in 2003. The Karnataka organic farming policy document has five section namely: farming for sustainability, principles and principal requirements of organic farming, objectives of the policy on organic farming, status of organic farming in the state, strategies for promotion of organic farming in the state. Here the first two are common for each state's organic farming policy and only remaining three sections are discussed.

Objectives: (a) To reduce the debt burden of farmers and to enable them to achieve sustenance and self-respect through organic farming. (b) To enhance the soil fertility and productivity by increasing life in soil. (c) To reduce the dependence of farmers for most of the inputs such as seeds, manures and plant protection materials by sourcing local natural resources thereby reducing the cost of cultivation. (d) Judicious use of precious water resources and maintenance of production level. (e) To improve farmer's income through production of quality produce. (f) To increase the food security by encouraging traditional crops and traditional food habits. (g) To increase the rural employment opportunities to prevent migration to urban areas. (h) To facilitate farmer's self-help groups for most of their requirements. (i) To make the environment safe and pollution free and to protect health of human beings and animals. (j) To equip the farmers to effectively mitigate the drought situation in rain-fed and drought prone areas. (k) To bring about suitable institutional changes in teaching and research on organic farming.

Status of Organic Farming in the State: The movement of organic farming taking place in Karnataka is not because farmers foresee a definite market for organically produced, but due to production-oriented reasons namely, reduction in use of external inputs, improvement of soil fertility, lower soil degradation, biological pest control and protecting mother earth besides improving their economy. A group of active farmers' associations involved in organic farming and state government are increasing awareness among the consumers about the organic foods. The state government recognized the importance of organic certification for exports and follows the Government of India's standards.

Strategies for Promotion of Organic Farming: The state has adopted integrated approach for the promotion of organic farming. (a)The state government have been working in one way or the other for promotion of organic farming and environmental protection activities in isolation in the state. (b) State-level empowered committee will setup for promotion of organic farming. (c) The Mini Mission on Organic Farming will act as an advisory body to the state-level committee for approval and technical support of organic farming and monitoring of projects. (d) Site-specific committees: During implementation of various programmes connected with organic farming in the state, the site-specific farmers' associations/farmers clubs/farmers' cooperatives/farmer's companies/SHGs/NGOs, with the prior permission form of the state-level committee that may set up another committee for preparation and implementation of action plans. (e) Area approach/commodity/crop approach and (f) Bio-mass production: Organic farming requires the production of biomass on farm. One cow per two acres will ensure the required compost production and cow urine on farm.

ZBNF: It is farming method which is rooted in the present movement and spread to various states in India. It has recorded huge success in the southern states of Karnataka and Andhra Pradesh. An estimate presented that around 1 lakh farm household are engaged in ZBNF in Karnataka. The movement on ZBNF is led by a farmer Subash Palekar in collaboration with Karnataka state farmer association, Karnataka Rajya Raitha Sangha (FAO, 2016). In India 296,438 farmers have committed suicide from 1995 to 2013 and debt was the one of the main reasons behind these suicides (Parvathamma, 2016). Under such vulnerable situation 'zero budget farming' promises to end the dependency on loans by drastically reducing the cost of production.

The term 'budget' refers to credit and expenses, thus the phrase 'zero budget' means without using any credit or spending money to purchase inputs and the term 'natural farming' means farming with nature without chemicals. ZBNF farmers are peasants and rooted with rural area villages. A survey conducted by La Via Campesina (LVC) in Karnataka found that ZBNF farmers have their own land, access to irrigation facility and at least one cow at home. The state government sponsored to organized state-level training camp in the state (FAO, 2016).

CENTRAL ZONE (MADHYA PRADESH)

Central Zone comprises two states of India namely Madhya Pradesh and Chhattisgarh. For the study of state organic farming policy Madhya Pradesh is chosen because it is the highest organic producing state and has a state organic farming policy. The Ministry of Farmers' Welfare and Agriculture Development, Government of Madhya Pradesh announced organic farming policy in 2010. The state occupies first position with having more than 1.48 lakh ha area under certified organic out of a total certified area of 3.40 lakh ha in the state. The organic farming policy of Madhya Pradesh intends to create, facilitate and strengthen the enabling environment for developing integrated value chains of the organic product for producers and consumers. The policy document invokes on 'farm to fork' approach to assure supply of 'healthy food for all'. The policy's long-term goal is to make Madhya Pradesh an organic state.

Policy Goals: The Madhya Pradesh organic farming document has three-fold goals considering cross sectoral, temporal and special factors which enhance the productivity. It has identified some threats like climate change, market unavailability and rainfall uncertainty, which may hamper to attain the policy goals. The policy document also invokes to enrich organic manure through biogas (*gobar gas*) for higher growth of organic farming in the state. The document has divided goals into short-term, medium term and long term. Short-term goal is to ensure enabling environment by developing capable and professional human resources and institutions necessary for both technology and market securities. Medium-term goal is to enhance return rates unit area under organic farming by rationalizing cost of production of the farm produce on one hand and increasing cash returns by augmenting market driven processes. And the long-term goal is to attain environmental sustainability through organic farming and management strategies leading to improved soil health.

Statutory Obligations: There are several statutory obligations in policy document namely organic certification, state-level agency, grower group certification, internal control system, chartered quality assurance managers, inspectors, ICS auditors, operators, PGS, level of equivalency with national and international standards and the policy stand against GM crops. The policy document also expects that all the statutory provisions which are mentioned above will help to promote organic farming in the state.

Organic Inputs: To assure supply of organic inputs at reasonable price for accelerating the growth of organic farming, the key inputs of organic farming are soil and plant nutrition supplements, plant protection agents, seeds and varieties of the crops. Some policy provisions are to facilitate mechanism for connecting bio-energy and organic inputs, to make the bio-energy organic input production commercially viable and sustainable, developing framework and augmenting opportunities for carbon trading towards indigenous co-based rural economy, gobar gas cleaning and bottling as domestic fuel, enriched manures with quality assurance, standards, organic input enterprises and rural youth and quality control facilities for inputs.

State Organic Mission: The organic farming policy would be implemented in such a way that it achieves a mission on organic farming that would be instituted to provide an umbrella organization institutionalizing effort for promoting organic farming in the state. The organic mission would be then implementing by an agency within the Ministry of Agriculture through a team of professionals that can manage all aspects of organic agriculture. The mission would work in all districts in the identified niches and shall develop full-scale operational facilities at district, block and cluster level. The mission would create an enabling environment in the state to encourage the organic producers by holding district and state-level competitions, institute awards and organize annual events such as regional, state, national, international growers' conferences, seminars, symposiums, workshops, exposure tours and *jaivik hats* for all the stakeholders engaged in organic farming.

Research and Development Initiatives: There are some policy provision which increase the research and extension services such as developing local, regional and state level facilities to impart short-term courses. For local level training of farmers, KVKs can be used as centres of organic farming. A state-level institute will be set up for research and development on organic farming. Policy document advocated to setup institutes for capacity building programme and assessment for all stakeholders to understand and spread the knowledge about organic farming. The state can step institution for training and earning of organic manure and preparation of field along with control of insects and research in organic farming.

Incentives: The following incentives are made for stakeholders by the state organic farming policy document. It initiates to set up a 'state organic farming fund' managed by state organic mission. The fund comes from regular budgetary allocations of farmers' welfare and agriculture development, a minimum of 20 per cent of the budget could be allocated to the activities approved by the state organic mission. Besides the above budget allocation, funds could also be generated through the Mandi Board cess tax regime and carbon credits could be utilized for extending grants and subsidies to popularize organic farming in the state. The compensation are available for only registered organic producers through subsidy of 25–75 per cent of the certification fee under individual farm certification, GGC and facilitate the free membership of PGS.

NORTH-EAST ZONE (SIKKIM)

North-east zone comprises eight states of India namely Arunachal Pradesh, Assam, Manipur, Meghalaya, Mizoram, Nagaland, Tripura and Sikkim. For the study of state organic farming policy, Sikkim is chosen because it was the first organic state in 2015 and formulated first organic farming policy in 2003. The state Sikkim has 729,900 hectares land out of which only 10.2 per cent (74,303 ha) is used in farming activities and rest of the area covered by forest. Sikkim is divided into five agro-climatic zones namely tropical zone, sub-tropical zone, temperate zone, sub-alpine zone and alpine zone. The agricultural land falls in tropical, sub-tropical and temperate zones. The state produces varieties of crops due to diverse agro-climatic conditions and state has fragile ecosystem for organic farming.

The fragile ecosystem in Sikkim hills demands sustainable farming practices which can protect natural resources and organic farming can save the natural resources by protection of the soil from degradation, conservation of ecology and healthy life for future generations. The state produces grains crops, horticulture crops and vegetables organically. The state organic farming policy has adopted four basic principles and given reasons for state adopted organic farming policy are as follows: state has low productivity with rainfed condition, farmers are traditionally growing crops organically and there is rich organic matter in soil.

Objectives: The government of Sikkim announced its organic farming policy through Gazette notification, dated 17 September 2003, notifying the constitution of the 'Sikkim State Organic Board' (SSOB) with the following objectives: (a) To address the basic requirement of organic crop production, wild crop harvesting, organic livestock management and processing and handling of organic agricultural products. (b) To develop state organic standards, listing out allowed synthetic and prohibited non-synthetic substances. (c) To set up accreditation authority for organic farming and foods in the state to administer and register state, national and international accreditation. (d) To establish an organic certification programme in the state specifying record keeping and labelling requirements. (e) To formulate organic regulations for producers and handlers in the state in relation to government, private and foreign organic certification programmes and in the interest of state. (f) To administer enforcement and appeal procedures to make sure all certified organic operations are in compliance with the national organic programme. (g) To make recommendations to the government to support organic schemes and programmes in the state. (h) To advise the government in creating a sustainable and competitive organic food and farming sector through appropriate action plans by identifying ways of achieving sustainable growth in organic farming, increasing share of market for organic produce through promotion and identifying measures required in the distribution, processing and retailing sectors to promote overall growth in the organic sector.

Implementation: The policy extended document has the following implementation provisions: ICS development through service providers, certification through APEDA accredited certification agencies, incentives for adoption of organic farming, on-farm production of inputs, off-farm certified inputs, large scale awareness and training programmes, market linkage development and branding with brand logo.

CONCLUSION

The constitutional provision provides the right to states to design their own agriculture policy due to agriculture being a subject of the state list. On organic farming the centre has specified the purity area, where the first to promote organic farming is the rain-fed area. So those states that have more rain-fed area are expected to promote organic farming and formulate their own organic farming policies. More than 17 states have formulated or drafted their policy documents after 2000. One state from each zone namely north zone—Himachal Pradesh, east zone—Odisha, central zone—Madhya Pradesh, west zone—Gujarat, south zone—Karnataka and north-east zone—Sikkim has been chosen for analysis of organic farming policy. All the state organic farming policies follow the four basic principles of organic farming and provisions of national organic farming policy regulations, along with additional state-specific goals.

REFERENCES

FAO (2016), 52 Profiles on Agroecology: Zero Budget Natural Farming in India, Case study provided by La Via Campesina, Accessed on June, 2018, http://www.fao.org/3/a-bl990e.pdf

Luttikholt, L. W. M. (2007). Principles of organic agriculture as formulated by the International Federation of Organic Agriculture Movements. NJAS *Wageningen Journal of Life Sciences, 54*(4), 347–360.

Ministry of Agriculture, Department of Agriculture & Cooperation. (2005). *Organic farming policy 2005*. New Delhi: Author.

Ministry of Farmers' Welfare and Agriculture Development Government of Madhya Pradesh (2010). *State policy on organic farming in Madhya Pradesh*. Author. Retrieved from http://www.mpkrishi.org/krishinet/hindisite/pdfs/javikneeti_eng.pdf

NCOF. (2005–2006 to 2013–2014). *National project on organic farming annual reports*. Ghaziabad: National Centre for Organic Farming.

Parvathamma, G. L. (2016). Farmers suicide and response of the Government in India: An Analysis. *IOSR Journal of Economics and Finance, 7*(3), 1–6.

ABOUT THE AUTHOR

Hari Ram Prajapati attained an MA in economics from the University of Allahabad and a doctorate in economics from the Central University of Gujarat. He qualified NET and received junior and senior research fellowship in economics for doctoral research by the UGC. His area of specialization in research is applied microeconomics with special reference to environmental economics, specifically the application of econometrics in consumer and producer behaviour analysis in adoption of organic farming, change in behaviour of sustainable consumption and production.

Dr Prajapati is working as an assistant professor in Economics Section, Mahila Mahavidylaya, Bananas Hindu University, Varanasi, UP. He is previously worked at the Department of Economics, Kamala Nehru College, University of Delhi (2016–2019). At present he is mainly engaged in teaching BA (Hons) Economics students. He is life member of the Indian Econometric Society and Indian Association for Research in National Income & Wealth, working for academic research in the area of applied econometrics and measurement of national income and wealth.

INDEX